Commercial Aviation in the Jet Era and the Systems that Make it Possible

T0172111

Thomas Filburn

Commercial Aviation in the Jet Era and the Systems that Make it Possible

Springer

Thomas Filburn
College of Engineering, Technology and Architecture
University of Hartford
West Hartford, CT, USA

ISBN 978-3-030-20113-5   ISBN 978-3-030-20111-1   (eBook)
https://doi.org/10.1007/978-3-030-20111-1

This Springer imprint is published by the registered company Springer Nature Switzerland AG
The registered company address is: Gewerbestrasse 11, 6330 Cham, Switzerland

To Max, Ella, and Mason, may you find as much happiness as you have brought into our lives. Kiki, thanks for everything.

Preface

This book contains a short history of commercial aviation beginning with the Wright brothers inaugural flight. It provides a more detailed glimpse into the aircraft subsystems like those found on the early piston engine propeller craft that flew the mail from Denver to Chicago. These same subsystems now permit large, turbofan-powered aircraft to fly intercontinental routes at speeds four times those early mail planes. The book explains the purpose of these subsystems and their evolution from flight's early days into the 500+ passenger aircraft available today. This book provides seven chapters dedicated to the design and operation of a multitude of subsystems required for large commercial aircraft to safely takeoff, cruise at Mach 0.85 (or higher), and land. It shows how these items and systems have changed from the early years of straight wings, open cockpits, and fixed landing gear to the swept wing, retractable landing gear craft that can connect us to the far reaches of the globe. The book includes seven additional chapters on the consequences of a component failure for each of the detailed subsystems. These components and design flaws frequently lead to the loss of aircraft which can be deadly for passengers and crew. This book demonstrates the complexity of today's commercial aircraft and the importance of the design, fabrication, and installation of these systems even those that operate for mere seconds at the end of each flight (thrust reversers).

West Hartford, CT, USA   Thomas Filburn

Acknowledgments

Thank you Taylor Goodell Benedum and Enrico Obst for the extra efforts you put in to make the graphics exemplary. Dan Patterson, thanks for the photos. Alan, I appreciate your editorial comments. Randi and the staff at the Mortensen library, wonderful support!

Contents

Nomenclature

ACM	Air cycle machine
APU	Auxiliary power unit
ATC	Air traffic control
Cg	Center of Gravity, imaginary point where the entire aircraft load is pressing downward
Cl	Center of Lift, a point where the entire lifting force is deemed to operate
CVR	Cockpit voice recorder
DFDR	Digital flight data recorder
ECS	Environmental control system
FAA	Federal aviation administration
FADEC	Full authority digital engine control
FAR	Federal aviation regulation
FBW	Fly-by-Wire
FDR	Flight data recorder
FOD	Foreign object damage
FPI	Fluorescent Dye Penetrant Inspection
GA	General aviation
GEAE	General Electric aircraft engines
GTF	Geared turbofan engine
LE	Leading edge
LWC	Liquid water content
mph	Miles Per Hour
MTOW	Maximum takeoff weight
NACA	National Advisory Committee for Aeronautics (precursor to NASA)
NASA	National Aeronautics and Space Administration
NEA	Nitrogen-enriched air
NTSB	National Transportation Safety Board
OBIGGS	On-board inert gas generating system
OEM	Original equipment manufacturer
PSA	Pressure swing adsorption

PSIA	Pounds Force Per Square Inch Absolute
QRH	Quick reference handbook
RTO	Rejected takeoff
SST	Supersonic transport
TE	Trailing edge
TR	Thrust reverser
TRL	Technology Readiness Level (1–9 developed by NASA to quantify concept (1) to flight demonstrator (9))
TSFC	Thrust-specific fuel consumption
USAF	US Air Force

Chapter 1
Commercial Aviation History

Early History

The traditional history of aviation points to the seminal event of the Wright Brothers powered flight at Kitty Hawk on December 17, 1903. Challenges to the Wrights as being first to powered flight arose on both sides of the Atlantic, including one Connecticut challenger Gustave Whitehead, but the aviation industry continues to credit the Wrights with the inaugural heavier-than-air flight [1]. The initial Wright flight only traveled 100 ft., but they were achieving distances of nearly 1000 ft. by the end of that December day, still at only an altitude of ~10 ft. [2].

Aviation's second decade showed a dramatic increase in capability, ostensibly spurred by the combat needs of World War I. While the initial Wright flight averaged less than 10 mph net speed, aircraft speeds increased to 125 mph near the outbreak of World War I [3]. This top speed had increased to 171 mph shortly after the end of the war [4]. This speed had more than tripled to 600 mph by the end of the Second World War [5]. While most of these top speeds were recorded by specialty or combat aircraft, they mirrored the increases seen in passenger aircraft (once that commercial enterprise started ~1920s). KLM, the airline having the longest run under its original name, started operating in 1919 [6].

The novelty of airplanes themselves provided the initial impetus for improvement in airplane capabilities. The Wright brothers spent the first 2 years after their invention developing the ability to control their craft. By fall 1905 they exceeded the endurance of all their previous flights, by completing over 30 min continuously airborne back at their home near Dayton Ohio [7]. Unfortunately, the Wright brothers were less inclined to improve their novel craft, instead spending a significant amount of energy defending their patent against primarily domestic challengers.

While the needs of the combatants in both world conflicts (WWI and WWII) clearly pushed airplane capabilities for speed, endurance, and load capacity, other commercial incentives spurred increases in aviation capacities as well. The first non-military endeavor in flight within the USA unmistakably stemmed from the

© Springer Nature Switzerland AG 2020
T. Filburn, *Commercial Aviation in the Jet Era and the Systems that Make it Possible*, https://doi.org/10.1007/978-3-030-20111-1_1

efforts of the US Postal Service to establish Air Mail Service [8]. This Air Mail amenity primarily relied on airplanes no longer needed by the Army Air Corps after WWI. This first decade of commercial aviation did not see large advances in airplane design, but the Air Mail service did spur development in ancillary aviation systems, such as airfields, direction finding, and methods to extend the range of the surplus military aircraft [9]. While commercial aviation commenced in the second decade of the twentieth century, passenger traffic did not provide a significant portion of the revenue for this market.

The maximum speed of aircraft had reached 184.3 mph by October 1920. Multi-engine airplanes were introduced during the First World War, but these were designed for delivering ordnance over a long distance, not passengers in comfort and safety. The 1925 Air Mail act, a response to corruption in the early airmail delivery funding scheme, also saw the real beginning of combined Air Mail and passenger service. Many historians point to the Ford Trimotor as a significant milestone in US commercial aviation. This aircraft, an attempt by Ford motorcars to find a new market, commercial air transport, introduced several new features. It had a completely enclosed cabin for passenger comfort. Henry Ford insisted on three motors for reliability, an important feature in 1926 with engines regularly becoming non-functional during in-flight operation. While the foray into airplane construction lasted less than one decade, it represented the first mass-produced model (199 fabricated), as well as other technological improvements. It introduced all-metal construction, a single cantilever (supported from the center fuselage, no struts providing support like in a biplane from the same era) wing. While a single wing added challenges like a higher wing loading (lift force/wing area), eliminating those support struts greatly reduced drag allowing for a faster cruising speed, which was further magnified by growth in engine power. The Trimotor was huge compared to other planes of the day, measuring over 50 ft. from nose to tail, with a passenger compartment that could hold 15 passengers [10].

By the end of the roaring 20s, airplane top speeds had increased to over 300 mph (357.7 mph, Supermarine Seaplane) [11]. The service ceiling and load carrying capability was also increasing during this era, during this short 4-year span, the Ford Trimotors no longer set the standard for passenger aircraft. Several new airplane designers and manufacturers began to ascend during this time. Jack Northrop eventually built a company with his name on the letterhead but would first work for the Lockheed and Douglas Aircraft companies. William Boeing started his namesake company in the Seattle, WA, area, and Donald Douglas was creating his self-named company in Santa Monica, CA.

Northrop created the Vega airplane in 1927 for the Lockheed (originally spelled Loughead) Corporation, the same year Lindbergh completed his solo Atlantic crossing. The single engine Vega would gain renown as the airplane used by Amelia Earhart in her headline grabbing flights and in her ill-fated attempt to circle the globe [8]. The six-passenger Vega incorporated a smooth circular fuselage and with its strong fuselage could support an engine of over 600 HP, allowing it to achieve a

top speed of 225 mph [11]. Wiley Post was another famous pilot who gained noto-
riety with the Vega and other planes.

Wiley Post would be remembered most for his pioneering work with pressurized suits to allow pilots or other crew members to work at high altitudes in unpressurized aircraft. Flying at higher altitudes with lower air densities reduced airplane
drag, allowing any plane to fly faster than it would at lower altitudes. However,
pilots and others were quickly realizing the limits on human endurance without
countermeasures as they reached these higher altitudes. Wiley produced and test-
flew a pressurized suit to allow him to breathe air at a higher density than the ambient low-pressure air inherent at high altitudes. The pressurized air allowed him to
remain conscious in his unpressurized aircraft, even manipulate controls and with
its umbilical allowed him to move around. Engine manufacturers also learned
about the limits of these higher altitudes on their air-breathing piston engines. They
soon adopted turbochargers and superchargers to improve the operation of their
engines in the rarefied air of high altitude. Both devices compressed incoming air
before its combustion in the cylinder, with one device (turbocharger) operating off
the high-temperature and high-pressure gases leaving the cylinder and the second
device (supercharger) compressing the incoming air directly via engine crankshaft
energy.

While it can be debated which airplane represents the first antecedent of the
modern airliner, the Boeing 247 represents a practical choice. The twin-engine, all-
metal airplane had a top speed of 200 mph, retractable landing gear, and could
accommodate ten passengers plus 400 lbs. of cargo (typically air mail) [12].

While it set a standard for capacity, speed, and technical advancement, the model
247 was soon displaced by the Douglas DC-2 barely a year after the Boeing model
was introduced. The DC-2 and its immediate successor, the DC-3, surpassed the
247 in passenger load (the DC-3 could sit 21 passengers), and would establish a
decades-long rivalry between the two airframe manufacturers. While 75 Boeing
247's were built, Douglas built 455 DC-3's along with over 10,000 of the military
cargo variant the C-47 [13].

Just before the USA's official entry into WWII, Boeing upped the ante in the
commercial airline market with the model 307 Stratoliner (Fig. 1.1). This four-
engine metal airplane looked similar to the B-17 bomber also being produced by
Boeing. In fact, the two variants shared wings, engines, and tail design. However,
the 307 operated with a pressurized fuselage (the B-17 did not). The pressurized
fuselage let the 33 passengers remain in at a pressure equivalent to 8000 ft. This
lower altitude (higher pressure) cabin provided for improved breathing and overall
passenger comfort, but did require a stronger fuselage to resist the pressure difference (designed for about 2.5 psi difference). The pressure difference allowed the
plane to routinely fly at 13,500 ft., and with its 20,000 ft. ceiling, it could reach
altitudes high enough to avoid many adverse weather conditions. At its nominal
cruising altitude, the cabin would be equivalent to flying at an 8000 ft. altitude
(slightly higher than Santa Fe NM).

Fig. 1.1 Boeing Stratoliner, first pressurized commercial Airliner 1939 (permission from the PanAm foundation)

Clipper Airplanes

Pan American Airways (PanAm) was the early US leader in International aviation travel and for many decades after it began. PanAm's ability to win US Air Mail contracts to International destinations helped it set the market standard for Caribbean, South American, and Pacific travel. Juan Trippe, the CEO of Pan Am for 50 years, displayed a strong interest in amphibian aircraft to support these International travel needs. His influence led to the introduction of the Sikorsky and Martin four-engine flying boats. These initial craft could house 40–50 passengers and had a range over 1000 nautical miles [14]. The amphibious nature of the craft provided flexibility for their island, and sea coast destinations. It allowed access to destinations without the expensive and difficult construction of airports and runways in areas without the capacity to build this type of infrastructure. Finally, the amphibious nature of the planes gave the illusion of safety for long transoceanic travel.

While the Sikorsky and Martin companies inaugurated this "Clipper" service, it was Boeing that produced the most memorable aircraft in this class. The Boeing models, also four engines, had a range of over 4000 miles, and could carry 75 conventionally seated passengers (40 in sleeper berths) [15]. But while the Clippers demonstrated utility and range, the exigencies of war would surpass their capabilities, and improved aviation infrastructure after the war would make them an anachronism in post-World War II aviation.

The first of the half of the 1940s saw aviation changes dominated by the needs of the combatants in the Second World War. These changes would ultimately provide benefits to the commercial aviation industry. It is interesting and coincidental that the jet engine would be developed by both Germany (Ohain) and Great Britain (Whittle) during the war, with Germany fielding several fighter plane models relying on jet engines. Many aviation experts point to the German dual engine Messerschmitt 262 as the best fighter of the war, it was powered by 2 Jumo 004 axial flow jet engines that could produce a top speed of nearly 560 mph, more than 100 mph above the P-51 Mustang or Spitfire, the premier Allied fighters at the end of the war [8]. The ME 262 also introduced wing sweep, a way to improve wing performance at the high subsonic velocities reached by airplanes near the end of the war (Mach 0.65–0.8) [Mach 1 velocity = speed of sound]. Figure 1.2 shows a later embodiment of the jet engine. While the Germans used an axial flow compressor, like that shown in Fig. 1.2, the British started with a centrifugal compressor which made for a broader inlet and ultimately higher drag. The Jet engine pulls in air, compresses it, raises its energy level in the combustion chamber, expands the gas in the turbine section, and expels the gas at high velocity out the back to generate thrust. The energy extracted in the turbine section is used to drive the compressor section, as the two sections are connected by a common shaft (or several common shafts).

The largest production (gross takeoff weight) airplane of the war was the Boeing B-29 Superfortress. This long-range bomber could weigh over 100,000 lb. when fully loaded, had a range greater than 3000 miles, a 30,000 ft. ceiling plus it provided a forward and aft pressurized crew compartment (vital at altitudes above 20,000 ft.). These range, ceiling, load, and pressurization features would set the standard for passenger aircraft after the war.

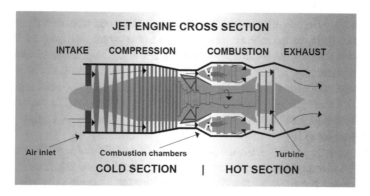

Fig. 1.2 Cross section of jet engine

Commercial Air Travel Post-World War II

At the end of the World War II, the public clamored for air travel, as it had been routinely denied to civilians because of the exigencies of the war. The airlines started converting numerous military cargo aircraft that were deemed excess by the Army Air Corps in 1945. American and United started with C-54 (the military version of the DC-4, a four-engine airliner that could carry over 40 passengers) cargo planes [16]. Passenger traffic ballooned too, commercial air traffic served 1.8 million US passengers in 1939 and reached 16.7 million passengers by 1949 [17], representing a 25% annual growth factor. Obviously, this growth rate was not uniformly seen during that decade.

While military surplus provided the initial capacity, air transport, airlines, and airliners continued to evolve in the 1950s. Lockheed introduced the Constellation, a propeller airliner with four piston engines that directly evolved from the US Army Air Force (USAAF) C-69 transport. The Connie, as it became known, could transport 100 passengers in pressurized comfort up to 4000 miles, certainly allowing transatlantic service [8]. Douglas advanced its DC-4, producing the DC-5 and DC-6 by 1947. The DC-6 once again set the standard for airliner features. This plane could fly 100 passengers at 300 mph, for 3000 unrefueled miles. Douglas built over 700 of these aircraft for civilian and military customers [18]. Boeing used its B-29 bomber as a model to produce 56 Stratocruisers, a 100-passenger airliner, with nearly ½ being bought by Pan American Airways [19]. These models would mark the end of piston engine propulsion for large commercial airliners. Starting in 1952 jet propulsion would transition from military aircraft to the civilian market. By this time these piston-powered commercial transport aircraft were cruising at an average speed of 300 mph (DC-6, Stratocruiser). The much higher cruising speed for the new jet aircraft meant a significant reduction in flight times between airport sites.

Commercial Jet Travel

It took nearly a decade after its introduction into military service for airframe manufacturers and airlines to embrace jet engines as a new propulsion method. It was the British who adopted the jet first employing them in the De Havilland Comet, first flown commercially by the British Overseas Airways Corporation (BOAC). The new turbojet propulsion provided a marked increase in top speed (480 mph vs. 300 mph from contemporary piston engine propeller craft) as well as cruising at a 40,000 ft. altitude, well above the 20,000 ft. limit for the piston engine propeller aircraft it replaced [8].

Unfortunately for the British, being first to market did not mean market dominance. The Comet had a major design flaw, its windows and the fuselage around them were not prepared for the fatigue caused by the multiple pressurization and depressurization cycles inherent in its operation. The Comet window had a generally square outline, allowing stresses to concentrate at the corners. The De Haviland Corporation had performed fatigue testing on a ground test airplane, but unfortunately this ground test article had first undergone an overpressure test that input a compressive stress into the fuselage. This compressive stress fooled the test engineers into thinking the plane could survive over 10,000 pressurization and depress events, ample enough cycles for the expected life of the aircraft. However, three Comets crashed between May 2, 1953 and April 8, 1954, with loss of all crew and passengers. BOAC grounded the entire fleet of Comets after the third crash, and the Airworthiness certificate was removed shortly after that third crash. It is ironic that the fatigue emanating from stress concentrations in a sharp corner would doom the Comet airplane. This same scenario (combined with brittle fracture) had caused many US designed and built Liberty ships to be lost during the early stages of World War II. The irony is that the British, one of many benefactors from Liberty ships, did not learn the lesson about stress concentrations in corner openings (Fig. 1.3).

De Haviland redesigned the Comet with round windows, a design that would prove appropriate for all pressurized aircraft designs to this day. Unfortunately for De Haviland, it would take until 1958 and the Comet 4 model, for De Haviland to have a solution that could meet regulatory requirements [8]. This would allow two

Fig. 1.3 De Haviland
Comet, square windows
and stress concentration

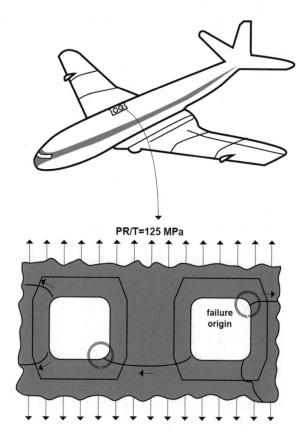

PR/T=125 MPa

failure
origin

airframe manufacturers in the USA to catch up and produce competing jet aircraft designs.

Boeing started its entry into commercial jet service with the Dash-80 prototype. This plane evolved into the 707, a four-engine jet aircraft that could fly at speeds of almost 600 mph, with 181 passengers, and at an altitude of 41,000 ft. It also demonstrated the benefits of the swept wing design, at the high speeds now possible with jet propulsion. The first flight of 707 occurred in 1957 and would start commercial service in 1958 with Pan American Airways. The 707 also put the jet engines in pods below the wings, a design option Boeing first attempted with the B-47 bomber program. It is noteworthy that this relying on pods slung under the wing would continue with their design to the present day for all Boeing and many competitor commercial jet airliners. This placement also helped with maintenance vs. the Comet design, which embedded the engines within the wing.

Douglas hoped its inaugural entry into commercial jet airplane design would continue its primacy in the commercial airliner market. Delta and United started flying the DC-8, a similarly sized passenger jet, relying on the same four Pratt & Whitney engines that propelled the 707. This model started service in September 1959.

As the jet era began in earnest in the USA, airline passenger traffic continued to grow at a prolific rate. US airlines carried 57.7 million passengers in 1960, representing a 345% growth in passengers since 1949. This equaled nearly a 12% annual growth rate since 1949, a rate much higher than GDP and population growth during the same period [17]. The benefits for jet propulsion (faster travel, reduced cabin noise and vibration) helped to overcome the increased fuel burn found in the jet propulsion. Some of this increased fuel burn came from the inefficiency of the early engine designs, while another cause was the higher drag produced by such a large increase in velocity vs. piston engines.

Boeing and Douglas both produced successful follow-on designs to their inaugural jet models. Boeing created the 727 model to expand jet service to smaller airports with shorter runways. The plane had three engines (all three tail mounted) and could transport up to 131 passengers [20]. It introduced several advancements into the commercial aviation industry including the auxiliary power unit. The APU is a smaller gas turbine engine that can provide electrical service to the aircraft, high-pressure air to the air-conditioning system to maintain cabin comfort without the main engines operating, plus high-pressure air to start the main engines. All these attributes would allow the 727 to become a popular and profitable model for Boeing with nearly 2000 airplanes produced by 1984 when the production ended.

Douglas introduced the DC-9 about this same time. This twin-engine (tail mounted vs. underwing) aircraft could fly 70–170 passengers in pressurized comfort for a range of about 1700 miles. It was the response of the Douglas aircraft to the perceived need for a shorter range aircraft similar to the Boeing 727. Nearly 1000 of these aircraft were built, including two slightly longer variants.

While Douglas and Boeing led the US commercial airframe manufacturing community in the USA, they had significant rivals. We already discussed the groundbreaking De Haviland Comet that suffered a design flaw from which it never recovered. However, France continued to innovate in the commercial aviation market, producing the short-mid range twin jet Caravelle in 1955. Sud Aviation produced nearly 300 of the Caravelle and became the first large-scale airframe manufacturer to house their engines near the tail. This engine location decreased cabin noise levels (except near the aft end).

After the success of the Caravelle, both French and British aviation companies began investigating supersonic transport. They expected that the next advance in commercial aviation aircraft would come from a greater than 2× increase in aircraft speed (1300 mph vs. the high subsonic velocities of the 1960s ~510 mph). Enterprises on both sides of the English Channel recognized the enormity of this challenge. This led to a cooperative venture between Sud Aviation and Bristol Aerospace (which became the British Aerospace Corporation (BAC)) to share the design and construction of the new aircraft, eventually named Concorde.

The Concorde was a bet on speed and the convenience of long distance intercontinental travel times being cut in half. In order to achieve the design goal of supersonic travel and overseas range, the team compromised on passenger load, with a normal complement being about 100 passengers, well below the 150+ passenger loads of the subsonic aircraft of the 1960s. The Concorde first flew as a prototype in 1969. Due to the complexity of the program and the many novel technologies required, it would require four prototype aircraft, and commercial service did not commence until 1976.

In the end only 14 Concordes would enter revenue service with Air France and British Airways, with each airline starting with seven aircraft. Unfortunately for these two airlines, the market for speed never materialized and the premium price required to pay for the high speed (low fuel economy) low passenger volume (100 passenger max) as well as limited airport service (noise restrictions and runway length) made for an unprofitable program. Due to resistance from environmental groups within the USA, which eventually killed the US Supersonic Transport program, the Concorde would be limited to scheduled service to only New York and Washington Dulles in the continental USA [21] (Fig. 1.4).

The very high aircraft speeds of the new craft led to some significant design changes between the SST and subsonic commercial aircraft from the 1960s through today. The delta-shaped wing is a hallmark of supersonic military aircraft, the engines are closely embedded below the wing, and the fuselage has a very narrow profile to minimize drag, which will also reduce the surface temperature rise that is inherent in these high-speed flights. The narrow fuselage allowed the available engine technology to push the aircraft to slightly greater than Mach 2 (2 times the speed of the sound), while limiting the skin temperature rise to manageable levels. However, this need for drag reduction would limit the aircraft to a single-aisle craft with room for only about 100 passengers. In addition to the drastic wing change vs. its subsonic brethren, the Concorde did not rely on horizontal surfaces in the tail for control. The high velocity of the aircraft meant that horizontal surfaces in this location would add significant drag, along with the potential for high pitching moments from control surfaces located this far aft.

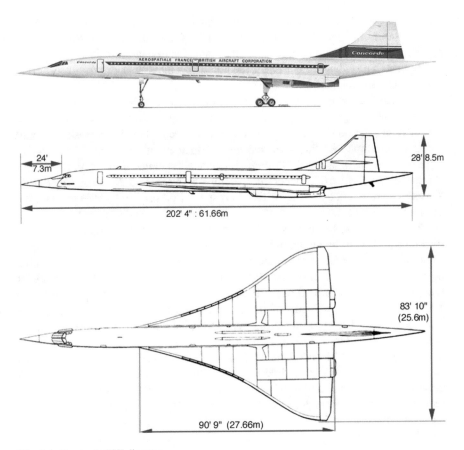

Fig. 1.4 Concorde SST diagram

The cooperation that led to the joint British-French Concorde also helped spur the pan-European aircraft consortium Airbus. Europeans recognized the dominance that American airplane designs and fabrication were achieving in the 1960s. They knew that left alone they would be simply subcontractors to the large American firms (e.g., Boeing, McDonnel-Douglas) unless they could achieve economies of scale and produce a European competitor to the large American firms. Hence Airbus was formed, initially with France, Germany, and the UK (37.5%, 25%, and 37.5%) starting the consortium. The Dutch and Spanish would be added to the mix soon after. The way the work was split, each partner would provide expertise for a certain component or subsystem. The British would supply the wings, the Germans, the forward and rear section of the fuselage, while the French built the center fuselage, cockpit, and control system. The French completed final assembly at the Toulouse site of the former Sud Aviation and other versions of the aircraft designer that had been nationalized.

What started as an endeavor to keep their aviation design and fabrication facilities relevant and operating has now grown to a leader in the worldwide aviation industry. Airbus and Boeing are now the remaining players in the large commercial airframe market with an ebb and flow of market leadership moving across the Atlantic Ocean. It seems that Airbus has taken the lead in the shorter-haul, narrow body (single aisle) segment with the adoption of the A320 neo family (new engine option, geared turbofan) exceeding the sales of its main competitor the venerable Boeing 737. Other players have emerged in the regional jet market, with Embraer and Bombardier offering smaller single-aisle aircraft. However, the wide-body market is still contentious and in fact it was Boeing's introduction of the 747 that started this wide-body or multi-aisle segment.

Wide Bodies

The US military recognized the need for more airlift capacity during the mid-1960s. Two factors strongly influenced this need, the ramp-up in the Vietnam War and the requirement to continue to support NATO allies in Europe vs. the Soviet Bloc threat. The US Air Force held a competition to develop a new heavy lift aircraft as a response to these significant airlift requirements. Lockheed won the competition in 1965 with their design of the C-5A Galaxy airplane. Boeing had entered the race and lost to Lockheed, this loss was the impetus for Boeing to then investigate a new design for commercial airliners. While Lockheed was busy designing and then producing the C-5A, US airline passenger traffic continued to expand at a fast pace. 150 million passengers were transported in the USA in 1968, a 260% growth since 1960, representing a 12% annual growth rate [22]. While not duplicated overseas, international travel and commercial air traffic in Europe was picking up at this time too.

Earlier responses to passenger growth and increased demand included new airfields and more scheduled departures, the airlines now anticipated meeting this continued growth with larger airplanes. In addition, they hoped that these larger planes would provide improved profits, making their seat-mile costs (the price to fly 1 seat, 1 mile) lower.

The final piece to create larger commercial aircraft was the ability to generate significant thrust with low-weight engines. GE's introduction of the high-bypass turbofan engine to power the new C-5 would provide that final technology leap allowing aircraft to leap beyond the capabilities of the then present generation aircraft such as the 727 and DC-9. Surprisingly, despite GE's early adoption for the C-5, Pratt & Whitney's high-bypass engines would power the initial 747 and all its variants.

The initial Boeing 747 model would employ four P&W JT9D engines capable of generating 43,000 pounds of thrust. The first prototype aircraft flew in 1969, with Pan Am taking delivery of the first production aircraft. Pan Am began revenue service with the aircraft on its New York to London route in January 1970. Dependent on the cabin mix of first class and economy seating, the plane could hold 374–490 passengers, a large increase in passenger load compared to the 130–170 passengers that could fit into the then common 727 and DC-9 aircraft (Fig. 1.5).

PAN AM HISTORICAL FOUNDATION

Fig. 1.5 747 picture (Pan Am historical foundation)

Shortly after Boeing's development of the 747, Douglas and Lockheed introduced competing models. Douglas made the DC-10, a three-engine aircraft with 2-5-2 seating arrangement (2 seats, aisle, 5 center seats, aisle, 2 seats). At its max capacity, the DC-10 could hold up to 380 passengers. The same model was produced for the Air Force (KC-10) as an aerial refueling tanker. Lockheed developed the L-1011, a 250 passenger, three-engine jet, with a very similar outline to the DC-10 (two wing-mounted engines, a third below the rudder). Unfortunately for Lockheed and Douglas, the wide-body market would not prove to be large enough to support four suppliers. While Douglas sold nearly 450 DC-10 models (60 as the military tanker), Lockheed would only sell 249 of its L-1011. Both airframe makers would be surpassed by the Boeing 747, which has delivered over 1500 models of the 747 in numerous variants (passenger, mixed and freighter version).

Airbus was slightly behind Boeing, and its other US competitors. The first flight test of its first wide-body, the A300 did not occur until October 1972, with commercial deliveries starting in 1974. The A300 aimed for the lower passenger end of the wide-body segment, initially sized for 250 passengers, and then quickly increased it to 270 passengers for the launch airline Air France. While late to the wide-body segment, Airbus has now added several wide bodies to the market, including the mega double-decker design, A380, which can hold 540 passengers in a mixed seating setup, but could hold 840 passengers in an all-economy configuration. A timeline containing some of the major milestones in the development of commercial aviation is shown in Fig. 1.6 below.

While it promised rapid connections around the globe, the supersonic transport caused significant public outcry in the USA. This opposition eventually led to its being cancelled as a design option by Boeing, which had an active government funded program until 1971, when government funding was dropped. The Europeans continued their supersonic aircraft effort, eventually fielding the joint UK-French Concorde. This consortium eventually built 20 of the supersonic airliners, of which

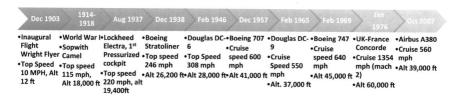

Dec 1903	1914-1918	Aug 1937	Dec 1938	Feb 1946	Dec 1957	Feb 1965	Feb 1969	Jan 1976	Oct 2007
•Inaugural Flight Wright Flyer	•World War I •Sopwith Camel	•Lockheed Electra, 1st Pressurized cockpit	•Boeing Stratoliner	•Douglas DC-6	•Boeing 707	•Douglas DC-9	•Boeing 747	•UK-France Concorde	•Airbus A380
•Top Speed 10 MPH, Alt 12 ft	•Top speed 115 mph, Alt 18,000 ft	•Top speed 220 mph, alt 19,400ft	•Top speed 246 mph •Alt 26,200 ft	•Top Speed 308 mph •Alt 28,000 ft	•Cruise speed 600 mph •Alt 41,000 ft	•Cruise Speed 550 mph •Alt. 37,000 ft	•Cruise speed 640 mph •Alt 45,000 ft	•Cruise 1354 mph (mach 2) •Alt 60,000 ft	•Cruise 560 mph •Alt 39,000 ft

Fig. 1.6 Notable milestones in commercial aviation

12 went into commercial service, starting in 1976. Its limited passenger load, high fuel burn, and an accident outside Charles De Gaulle airport in 2000 contributed to the eventual retirement of this aircraft type.

The following chapters will detail the changes to the aircraft subsystems that have been introduced during the jet era. The low-drag wing geometries required for higher speed jet travel need special augmentation to generate significant lift at the low speeds required for takeoff and landing. In addition, how to get a one-million-pound aircraft to change direction is not trivial, so a chapter on flight control is included.

Jet engines have changed over the decades, from the original pure jet design to the high-bypass flow machines of today. These changes have allowed designers to incorporate new features such as noise attenuation and thrust-reversing mechanisms (for reduced brake wear upon landing).

While jet travel did not initiate high-altitude pressurized flight, it has made it a reality for millions of travelers. We will provide a look at the systems that keep the cabin pressurized to a reasonable level, while keeping it comfortable despite the −50 F air temperature that can be found outside the fuselage.

Frequently ignored in airplane travel, but important for both takeoff and landing, are the landing gear and wheels used. The much higher landing and takeoff speeds of the initial turbojet and now turbofan engines have put much higher loads on the wheels, brakes, and landing gear of these aircraft. While only used during two, brief, phases of flight, their importance cannot be overlooked.

How do we store enough fuel to get a passenger plane from LA to Beijing? The method of where to store it, how to safely transfer it to the engines is documented. The high-pressure environment of the engine, coupled with the very low ambient pressure and temperature of the flight regime make for a significant pumping task to keep the engines running at high power, with high-temperature air. This fuel flow task can be made more difficult by flight maneuvers which will add acceleration loads in direction away from the gravity vector.

How do pilots know where they are heading, how high they are, and what speed they are traveling is also included. Unlike ground vehicles, airplanes do not have the luxury of wheels in constant contact with the ground for documenting speed and distance traveled. While GPS has provided improvements, pitot probes and Angle of Attack indicators that were introduced in the 1920s and 1930s remain important instruments for today's commercial aircraft.

The final subsystem chapter will discuss the measures incorporated to keep ice and lightning from affecting the aircraft. This chapter will discuss the designs and methods used to shed ice periodically before it gets too heavy (deicing) or to prevent its deposition at all (anti-ice). In addition, this chapter will discuss the methods used to keep aircraft safe and stable while traveling through lightning environments, where voltage discharges can exceed ten million volts.

Chapters 9–15 will detail events that have happened when each of the subsystems do not perform as designed. In some cases, the fault is pilot error or training, in others the fault lies with inadequate design, manufacture, or assembly. However, the results of these subsystem failures have caused fatalities and can lead to a total loss of the aircraft together with the crew and passengers.

References

1. Air & Space Magazine, Sep 2013, Who Flew First, Tom Crouch
2. National Park Service, https://www.nps.gov/wrbr/learn/historyculture/thefirstflight.htm, visited 7/15/17
3. http://www.earlyaviators.com/eprevost.htm
4. Munson, Kenneth (1978). Jane's Pocket Book of Record-breaking Aircraft (First Collier Books Edition 1981 ed.). New York, New York, USA: Macmillan. ISBN 0–02–080630-2
5. British Fighter since 1912, Naval Institute Press; First Edition (April 1993
6. http://www.burnsmcd.com/insightsnews/insights/aviation-special-report/2011/timeline-of-commercial-aviation, retrieved July 17, 2017
7. American Aviation Heritage, March 2011
8. Airplanes, the Life Story of a Technology, J. R. Kinney, Smithsonian Institution, 2006
9. The Saga of the Tin Goose: The Story of the Ford Tri-Motor 3rd Edition 2012, David Weiss, Trafford Publishing
10. Supermarine Aircraft since 1914, London:Putnam, 1987, Andrews, C.F. and E.B. Morgan, ISBN 0–85177–800-3. alt. Jane's pocket book of Record Breaking Aircraft, Edited by John Taylor, Collier Books, 1978, NY NY
11. http://www.lockheedmartin.com/us/100years/stories/vega.html retrieved July 23, 2017
12. http://www.boeing.com/history/products/model-247-c-73.page retrieved July 24, 2017
13. http://www.boeing.com/history/products/dc-3.page retrieved July 24, 2017
14. http://www.centennialofflight.net/essay/Commercial_Aviation/Pan_Am/Tran12.htm accessed Aug 12, 2017
15. Boeing, the First Century, Bauer, E. E., 2000, TABA publishing
16. http://www.boeing.com/history/products/dc-4.page retrieved July 24 2017
17. Facts and Figures about Air Transportation, 22nd edition, 1961, Air Transport Association
18. http://www.boeing.com/history/products/dc-6.page retrieved July 26, 2017
19. http://www.boeing.com/history/products/model-377-stratocruiser.page retrieved July 26, 2017
20. http://www.deltamuseum.org/exhibits/delta-history/aircraft-by-type/jet/boeing-727 retrieved July 20, 2017
21. http://www.concordesst.com/techspec.html retrieved July 30, 2017
22. https://www.rita.dot.gov/bts/sites/rita.dot.gov.bts/files/subject_areas/airline_information/air_carrier_traffic_statistics/airtraffic/annual/1954_1980.html retrieved 8/4/17

Part I
Sub-systems

Chapter 2
Flight Controls, High-Lift Systems, and Their Actuation

Fundamental Control

The present generation of commercial airliners uses a tail that controls trim and directional stability. Directional stability represents control over the airplane's direction, while trim refers to the instability (torque) generated by the center of lift (the large upward force keeping the airplane aloft) acting through a different point than the center of gravity. This difference in forces means that the aircraft will have a constant torque that must be counteracted, or the airplane will twist out of control (Fig. 2.1).

It was the Wright flyer that first demonstrated a practical approach to providing that aerodynamic control, but they located these control functions in front of the main wings in their (and the world's) first heavier-than-air powered aircraft. It would only take a few years for the Wrights to move these same control features from forward of the wings to aft. These aft-mounted control surfaces would remain behind the main wings for the remainder of the twentieth century and continue today in all commercial aircraft. Aircraft design started and predominantly remains with the main wings close to the center (fore and aft) of the aircraft, inherent to this symmetric design, it will place the wings in the center from a port-starboard view. The wings provide the dominant contribution to aircraft lift. The engines must overcome drag from all the surfaces that are directly in-line with flight, and skin friction along their entire surface. The drag force that the engines overcome strongly depends on aircraft speed. As a function of the forward velocity squared, it will generate a 4× increase in drag for a doubling of speed (identical aircraft). This drag force proportional to velocity squared and many other technical and financial hurdles (e.g., noise, limited payload, high fuel cost, high development cost) have limited present-day commercial aircraft to the subsonic regime (Fig. 2.2).

The general design of most World War I combat aircraft relied on two parallel wings (Biplane) with a rear-mounted tail for directional and pitch control. The pilot now seated (vs. the Wright flyer prone position) had a control stick and rudder

© Springer Nature Switzerland AG 2020
T. Filburn, *Commercial Aviation in the Jet Era and the Systems that Make it Possible*, https://doi.org/10.1007/978-3-030-20111-1_2

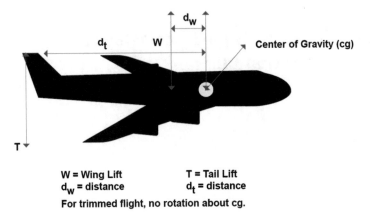

W = Wing Lift   **T = Tail Lift**
d$_w$ = distance   **d$_t$ = distance**
For trimmed flight, no rotation about cg.

Fig. 2.1 Forces acting on aircraft in flight

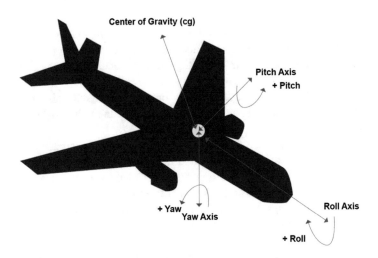

Fig. 2.2 Fundamental control axis for airplane stability

pedals for control. This control stick and rudder design would stay constant for many decades, and it was not until Airbus introduced the sidestick controller that it varied for a significant segment of the commercial aircraft market.

The control stick (also called a yoke) could be pulled back (or pushed forward) and this action moved control cables from the cockpit to the tail. These cables combined with a pulley system would elevate (control stick movement) or depress the (normally) horizontal elevator surfaces at the rear of the tail surface. The aircraft response would be nose up (elevator up, control stick pulled back) or nose down (elevator down, control stick pushed forward). The airplane would climb or move lower based on this control surface movement.

Fig. 2.3 DC-3 cockpit
with control yoke and
rudder pedals. (Source
Mid-Atlantic Air Museum)

In addition to the forward and backward motion of the control stick, the yoke (control wheel shown in Fig. 2.3) could be rotated left or right (clockwise or counterclockwise). This rotation, intended to start the aircraft banking into a turn, would raise the aileron on one wing and lower the aileron on the opposite wing (control surfaces at outer edge of wings). These actions would tend to roll the aircraft about its longitudinal axis. In addition, the action would tend to yaw the aircraft (rotate about central axis looking down on aircraft). Unfortunately, the yaw would be opposite the intended direction of turn. Therefore, using the rudder pedals, the pilot will push one pedal to direct the rudder, this action will yaw the aircraft in the proper direction, both overcoming the adverse yaw from the aileron movement and adding more rotation in the intended direction [1].

Control Surface Actuation

Designers of the first aircraft used simple cable linkages to translate the control stick and rudder movement into deflections of the ailerons, rudder, or elevator. These first generation aircraft were small and slow. The small size meant that small flight control surfaces could generate significant changes in the aircraft flight direction. Slow speeds meant that the aerodynamic load on the flight control surfaces were also low, allowing a pilot to move these surfaces with their own muscle power. A small iteration on this design used mechanical linkages, with an offset to the control surface and cable sheave. This offset allowed the pilot to increase the force applied to the control surface beyond what was introduced into the control stick. Figure 2.4 shows a photo with the canvas skin removed from a SE 5A biplane from WWI with a crank that helped move cables and ultimately the aircraft control surfaces.

Figure 2.5 shows a schematic of how this crank and pulley arrangement could be arranged to achieve the early flight control actuation.

Fig. 2.4 Interior flight
control actuation assembly,
SE 5A biplane (copyright
Dan Patterson, by
permission)

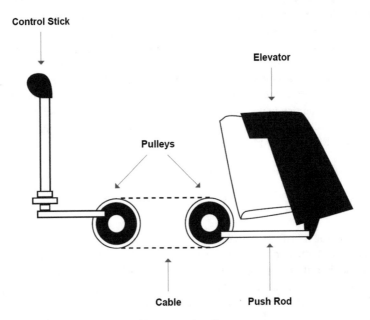

Fig. 2.5 Mechanical linkage to move flight control surface

These simple mechanical connections provided sufficient force during World
War I and through the interwar years. As aircraft speeds increased along with aircraft
size at the start of the Second World War, this simple mechanism could no longer
provide the forces necessary to effect flight control surface movement. In the 1930s,
NACA investigated the control forces that could be generated with the stick and

yoke/wheel assembly. They determined a maximum force of 35 lbs. into the stick for longitudinal control and 80 lbs. into the wheel rim for lateral control [2].

To overcome the inadequacies of pure muscle power and mechanical linkages, designers began to introduce hydraulically boosted systems to move the actuator and eventually the control surface. These systems relied on mechanical linkages to move control valves, which then allowed hydraulic fluid to travel to an actuator located at the control surface. This hydraulic fluid was at a high enough pressure (3000 psi by World War II) and had sufficient area to overcome the higher aerodynamic loads that were opposing the control surface movement [3]. A significant advantage to the pilot from the early direct control setup came from the immediate feedback from the aerodynamic forces on the control surface, through the same linkage, into the control stick. This feedback tended to be reduced or even eliminated in the hydraulic actuation systems.

As aircraft size and speed increased so did the energy that needed to be dissipated during the landing process. The very early aircraft relied on the drag created by the ground to slow the plane down. However, by the 1920s airplanes were large enough and landing at faster speeds such that this gradual process would not provide sufficient retarding force to allow planes to slow down in a reasonable distance. Drum and Disc brakes were already being used for the automotive industry and so they were also incorporated into aircraft, using the same pedals that operated the tail surface to apply a braking force. The larger mass and higher speeds of the aircraft necessitated a system to augment the pilot's foot braking movement. This need produced the first hydraulic system on airplanes, hydraulic pressure to activate the brake surfaces.

The hydromechanical flight control system began its introduction during the Second World War. These systems kept the same yoke and rudder pedals that were found in most cockpits, but augmented the pilot's force via a hydraulic fluid circuit. As seen in Fig. 2.5 below, the same yoke assembly is mounted in the cockpit. However, the output of the pilot's action on the yoke does not directly translate to the flight control surface. A movement of the control stick will produce a movement in the control valves of the hydraulic circuit. Small movements will generate small openings in the control valves. These small valve openings will allow limited hydraulic pressure to reach the control surface. Larger control stick movements will open the control valves to a greater degree, allowing higher hydraulic pressures to reach the control surface actuator (Fig. 2.6).

The benefit of the hydromechanical flight control system was generally introduced on larger aircraft during World War II, e.g., the Merlin Lancaster (four-engine British bomber that could weigh up to 55,000 lb.) used this system to control the flaps on the main wings during various phases of flight. However, Lockheed used the same system to boost the actuation of the ailerons on its P-38 twin-engine fighter (17,000 lb. weight). The large (for a fighter) airplane was difficult to roll due to its large wing area and engine placement. The hydromechanical system allowed the ailerons to be fully deployed using 1/6 the force of the purely mechanical system [4]. Douglas aircraft tested hydraulically boosted flight control actuation on a

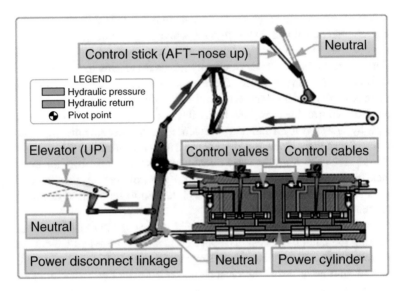

Fig. 2.6 Hydromechanical flight control system

prototype DC-4 aircraft but rejected incorporating them into their production aircraft. Boeing delivered this boost system on their Model 307 Stratoliner [5] and it became the first commercial airliner to use this control mechanism.

High-Lift Devices

Airplane designers needed to find new ways to transition from landing or takeoff configurations to cruise as the airplane speeds and wing loadings (lifting force over wing area) increased. The higher aircraft cruising and top speeds generated higher lift for the aircraft, but these same aircraft needed a method to generate high lift at lower aircraft speeds during takeoff and landing.

High-lift devices (flaps and slats) began to be introduced in the 1930s. These devices provided enhanced wing lift at the lower speeds required for takeoff and landing. Lower landing and takeoff speeds were not required by early aircraft as their top speed and cruising speed were reasonably slow enough, that their landing and takeoff speeds would not be excessive. As aircraft speeds increased, initially their takeoff and landing speeds saw a commensurate increase. It would only take until the 1930s for aircraft speeds to reach a magnitude that created difficulty in maintaining reasonable takeoff and landing distances. Lower takeoff speeds meant a shorter takeoff roll, allowing runways to be used without extraordinary distances. Presently, major hub airports (e.g., Hartsfield in Atlanta and O'Hare in Chicago) have runways that reach 12,300–13,000 ft. long. Secondary market airports or those with significant land constraints are in the 7150–9500 foot range (TF Green,

Providence RI, Logan, Boston MA, Bradley Hartford CT). Due to the increased lift force required at takeoff inherent with their large fuel load, takeoff velocities usually surpass landing velocities on an individual aircraft type.

Lower landing speeds has the benefit of shorter stopping distances and significantly lower braking energies (Brake energy is a function of landing speed squared). Lower landing speeds also translates to lower wear on aircraft tires, which must go from zero rotation to landing speeds of 150+ mph. Another significant consideration in designing aircraft is to maintain a significant delta between the landing speed and stall speed. Civilian transport aircraft typically aim for a landing speed at least 20% greater than their stall velocity (the velocity at which the lift dramatically drops) [6]. This 1.2 factor allows for sufficient variability in landing conditions, to preclude a catastrophic condition like stall.

The DC-2, the model which started Douglas Aircraft's early dominance in the commercial airline arena, appears to be the first airliner to incorporate a trailing edge flap (split) to increase wing lift and reduce landing speeds [7]. Figure 2.7 below shows the split flap used by the DC-3 and many other aircraft from the 1930s to the 1950s [6]. Due to the increased drag from the split flap, aircraft have evolved to slotted flaps to provide increased lift, with less drag than these early split flap designs. The advantage of the leading edge and trailing high-lift devices in Fig. 2.7 are their ability to be deployed for brief iterations (e.g., takeoff or landing) for the increased lift needed during these important low-speed operations. In addition, these same devices can then be stowed for reduced drag during higher speed and

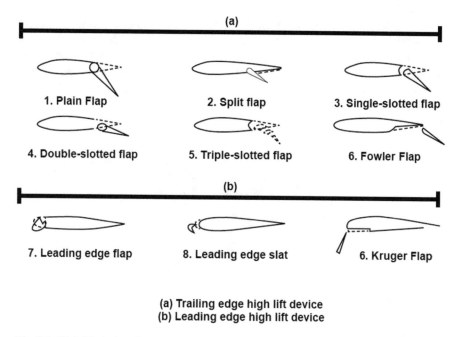

(a) Trailing edge high lift device
(b) Leading edge high lift device

Fig. 2.7 High-lift devices for commercial aircraft [8, 10]

inherently higher lift conditions (cruise) when the devices and their increased lift (but inherent higher drag) are no longer required.

A modified de Haviland DH4 reconfigured to a monoplane (originally designed and built as a biplane) was one of the first aircraft to test these high-lift devices. It employed leading edge slats and trailing edge flaps to improve the low-speed performance. These devices were imperative as this low-speed lift was lost when it was converted to a monoplane [9]. The devices shown in Fig. 2.7 were developed from the 1930s through the 1960s. The trailing edge flaps changed from the simple but high drag plain flap to the split and Fowler variant, which provided higher lift without any great difficulty. The slotted flaps (single, double, and triple) actually increased the wing length, thereby providing two methods of increased lift (greater wing area and higher lift due to flap position/actuation). The multitrack Fowler-type flap can be found on the two remaining large commercial aircraft designer/manufacturers (Airbus and Boeing) [9]. The devices at the leading edge also increased lift and for designs seven and eight in Fig. 2.7 provided increased wing area. The leading edge devices also increased drag and so identical to the trailing edge devices, they will be stowed after takeoff to decrease fuel burn during cruise when their increased lift is not needed [6].

Figure 2.8 shows all the flight control surfaces used on modern jet transport aircraft. The rudder on the vertical tail controls movement about the yaw axis (see Fig. 2.2). The elevators control pitching movement (previously described). The ailerons control rolling movements. As aircraft have gotten larger, with increasing wing length, spoilers have been added to the top surface of the wing to aid aileron roll control. When raised above the wing these devices spoil the wing lift (hence the name) and help roll the aircraft in the desired direction (e.g., spoilers on the wing to be lowered will be activated). They can also be activated in tandem on both sets of wings to act as in-flight speed brakes. Finally, they are frequently activated upon landing to rapidly lose lift and thereby increase the vehicle weight on the wheels. Braking forces can be increased when the entire aircraft weight is loaded onto the landing gear and wheel assemblies [11].

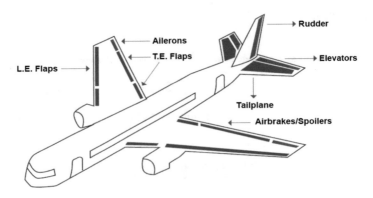

Fig. 2.8 Typical jet transport flight control surfaces and locations

The system for moving these various flight control surfaces has evolved over the decades. As described earlier, the first mechanisms relied on cables and pulleys to move the elevator, rudder, and ailerons (primary flight control surfaces) on the aircraft. The primary flight control surfaces are so named because of their criticality in maintaining controlled flight. Secondary flight control surfaces (flaps and slats, high-lift devices) are so named because their inadvertent actuation would not produce loss-of-control of the aircraft.

The load required to move the primary flight control surfaces increased as the aircraft size and weight increased. It became untenable to rely on pilot exertion, even with mechanical leverage to move these surfaces with the loads generated by the size and velocity of multi-engine World War II aircraft. The solution came from extending the hydraulic system used to brake the aircraft on landing, into a flight control actuation system. The early systems for controlling actuators used hydraulic systems operating under 1500 PSIG (pounds force per square inch, gauge).

By the 1950s, military and commercial aircraft were relying on hydraulic actuators to move their flight control surfaces, stow and deploy landing gear, and apply brake force. Early examples of fully hydraulic aircraft include the P-80 Shooting Star (first US jet fighter) and the Lockheed Constellation passenger airplane. Also in the 1950s, the hydraulic systems upgraded to 3000 psig systems. Newer tubing alloys and improved connections allowed for hydraulic systems to operate reliably at these higher pressures. Higher operating pressure allowed the end effector and the operating piston to be smaller, thereby decreasing the size and mass of these devices on the airplane. However, by the twenty-first century even higher loadings were required for aircraft as diverse as the V-22 and the A-380. Both of these aircraft rely on hydraulic actuators that operate at 5000 psig. Higher operating pressures are possible, but thicker walled hydraulic lines, heavier more powerful hydraulic pumps, and difficulty with seals at higher operating pressures make this option less desirable for most aircraft design groups.

While higher hydraulic pressures will provide stronger actuation force (for the same actuator area), other hydraulic problems forced researchers into a search for improved hydraulic fluids. Water is not a good fluid because of relatively high freezing point among its other detractions. Early hydraulic systems used a vegetable oil-based fluid, and as previously mentioned the hydraulic system started as a method to apply a large force for braking. Brakes, especially aircraft brakes, get very hot. The proximity of the flammable vegetable oil-based hydraulic fluid to the hot brakes led to numerous vehicle fires and the search for a fluid with similar properties (e.g., viscosity, freeze point, boiling point), but much lower flammability. Skydrol is a brand of phosphate ester hydraulic fluids that has evolved since its first introduction in 1948 in response to this need for reduced flammability [12].

Fly-by-wire systems began in the 1970s as a lighter weight, more responsive system for positioning of flight control surfaces that also offered reduced maintenance. The fly-by-wire system replaces the mechanical linkages, with local force transducers and a computer. It retains the hydraulic or electromechanical actuator at the flight control surface. Instead of direct pilot action being turned into control surface movement, the pilot's actions (rudder, yoke, or sidestick control movement)

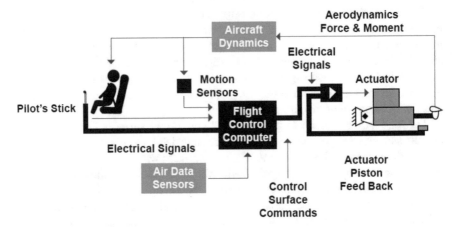

Fig. 2.9 Fly-by-wire flight control system

are turned into actuation signals by force and motion transducers in the cockpit. These signals are compiled by the flight control computer which then directs flight control surface actuators based on these pilot movements. The flight control computer also has motion sensors to confirm the aircraft response based on the pilot inputs (close the control loop). Figure 2.9 shows the embodiment and major components of a FBW system. While initiated in military aircraft in the 1970s, almost all civilian airliners rely on this control scheme today. It can be susceptible to electromagnetic interference (EMI) and high-intensity radiated fields (HIRF). Shielding can protect the wires and control computers from these potential interference sources but adds significant weight to the system [13].

Fly-by-*light* has been proposed as the next improvement in aircraft flight control systems. This system will be similar to fly-by-wire in that pilot movements are translated into signals from force and motion transducers. However, the transducer to computer signals and flight control computer to actuator signals will travel on fiber optic cables. This system will be much more resistant to EMI and HIRF attack, as well as offering decreases in system weight.

All of the later flight control schemes listed use redundant cables and channels to provide safety on commercial aircraft. The hydromechanical and fly-by-wire systems can use three or four channels to provide backup in case the primary channel is compromised. It was difficult to incorporate multiple backups to the direct mechanically linked systems used on the early flight control actuation systems.

Flight Control Surface Actuation

Figure 2.4 shows an example of how mechanically linked systems can actuate movement in a flight control surface (tail elevator). In this example movement of the control stick (fore or aft) produces movement in a pulley, which generates cable movement and therefore rotation of a second pulley. A control rod connected to the second pulley will force a short lever up or down around the hinge point of the elevator. The small surface area of the elevator forward of the hinge point has been added to reduce the aerodynamic load on the elevator and thereby reduce the load on the mechanical actuation system. This arrangement allowed pilots to continue to use mechanical control surface actuation, when the overall surface aerodynamic loads started to exceed the limits of their muscle power. Not shown in this figure, but a second control yoke can be added to the mechanical mechanism, allowing a copilot to also fly and input flight control movement, as well as flight control power.

Figure 2.5 shows the capabilities and the complexity introduced by hydromechanical flight control actuation. In this incarnation, control stick movement will directly affect the position of control valves connected to the hydraulic system. Movement of the control valve will meter higher pressure hydraulic fluid into one side or the other of a power cylinder. The difference in pressure on each side of the power cylinder will drive it in one direction or the reverse. This power cylinder movement is directly coupled to the flight control surface, again raising or lowering the surface as directed by input from the pilot movement.

Additional actuation devices, beyond the simple hydraulically moved cylinders exist for the numerous flight control surfaces. For takeoff and landing the leading edge and trailing edge devices will be operated in tandem by hydraulically powered motors. These motors then drive torque tubes that rotate geared actuators on both the leading edge and trailing edge devices.

The pitch control elevators typically rely on hydraulically powered motors that drive a jackscrew. A jackscrew is a long-threaded rod that can transfer rotary motion into a variable linear adjustment. This allows the elevator position to be adjusted by a rotary input. The rudder control is actuated by traditional hydraulic actuators, changing the rudder position by increased or reduced force on the actuator cylinder.

Not a control device, but wing tip additions are now visible on many transport aircraft. The winglet (upward turning appendage) and the scimitar (T-shaped edge effect) can be seen on the wing tips of many commercial and military transport aircraft (C-17, 737, A320). Both designs reduce fuel burn by taking advantage of the wing tip vortices that are generated in conventional wing designs. The pressure difference between the lower and upper wing surface provide the fundamental lift for the aircraft. However, this same pressure difference generates vortices (that can be seen in certain air-conditions as rotational flows which trail from the wingtips) that add drag to the airplane. The new wing tip devices convert those vortices into additional thrust, enough to overcome the added weight and surface drag of the new devices.

In summary, new high-lift devices had to be developed for the jet aircraft. These high-speed aircraft had their wing designs optimized for the cruise phase of their flight regime. However, these same wings could not provide sufficient lift at the much lower speeds required for takeoff and landing, hence they incorporate flaps and slats. These devices, only used during takeoff and landing, increase the wing lift during these lower speed operations. A second change resulted from the much higher aerodynamic loads that are placed on control surfaces when operating at $M = 0.85$, vs. the lower speed propeller driven velocities. New systems for movement and sensing were required to provide adequate aircraft control.

References

 1. Pilot's Handbook of Aeronautical Knowledge, FAA-H-8083-25B, 2016
 2. Airplane Stability and Control, Abzug, M., Larrabee, E., 2nd Ed. Cambridge University Press, 2002, ISBN 0 521 80992 4
 3. The Evolution Of Flight Control Systems Technology Development, System Architecture And Operation, Al-Lami, H., Aslam, A., Quigley, T., Lewis, J. Mercer, R., Shukla, P.
 4. US Patent US 2389274 A "Aircraft Control System", Pearsall, 1945
 5. Aviation Week & Space Technology, May 9–22, 2016, p67
 6. Aircraft Design, A systems Engineering Approach, Sadraey, M., 2013, John Wiley and Sons, ISBN 9781119953401
 7. Diamond Jubilee of Powered Flight, The Evolution of Aircraft Design, Pinson, J. Ed., 1978, AIAA
 8. A History of Aerodynamics: And Its Impact on Flying Machines, Anderson, J., Cambridge University Press, 1999, ISBN 9781139935999
 9. Aviation Week & Space Technology, May 9–22, 2016, p70
10. High Lift Systems on Commercial Subsonic Airliners, NASA CR 4746, Rudolph, P. Sep. 1996
11. Aircraft Flight Control Actuation System Design, Raymond, E., Chenoweth, C., SAE 1993. ISBN 1-56091-376-2
12. http://aviationweek.com/improving-fleet-performance/skydrol-history-aircraft-hydraulics retrieved Sep 7, 2017
13. Evolution of Aircraft Flight Control System and Fly-By-Light Flight Control System, Garg, A., Linda, R., Chowdhury, T., International Journal of Emerging Technology and Advanced Engineering, vol. 3, Issue 12, Dec. 2013

Chapter 3
Engines and Nacelles

Thrust Reverser

The gas turbine (turbojet or jet) engine began to supplant the piston engine propeller commercial aircraft in the 1950s. By the 1960s the jet engine was the dominant propulsion method on most commercial long-haul routes. A hybrid, a jet engine direct driving a propeller retained some advantages in short-haul routes. The reason for the jet domination can be attributed to several attributes of the jet engine vs. the piston-propeller combination. The jet engine provided greater thrust per pound of engine mass, thereby providing a means of achieving higher aircraft speeds. Figure 3.1 demonstrates this trend showing the improvement in the weight-to-power ratio from the reciprocating engine through the gas turbine years.

The jet engine represented a new type of prime mover, completely different from the piston engine craft that had been the hallmark of airplane propulsion for the first 50 years of powered flight. The sections of the jet engine include the inlet, in which air enters at the speed of the aircraft, the compressor section where its pressure is increased, the combustor region in which fuel is burned to greatly increase the energy available in the air, the turbine section which extracts some of the energy from the high-temperature air to drive the compressor region, and finally the exhaust in which high-temperature gas leaves the engine at a high velocity to create the forward thrust. The first jet engines by both the Germans and the British used a single shaft to connect their drive turbine to the compressor section. Figure 3.2 shows schematically how these sections can be arranged in a jet engine.

In order to avoid the large instabilities that come from shock waves, propeller aircraft need to keep their blade tips rotating below the speed of sound (blade velocity is blade length times rotational speed, hence the highest velocity is at the tip). This together with the limits in power density (torque produced per time divided by engine weight) of piston engines meant that the piston propeller combination was nearing its limits by the end of the Second World War. Figure 3.3 shows the trend in commercial aviation cruise speed, with the DC-7 being one of the last of the large

© Springer Nature Switzerland AG 2020
T. Filburn, *Commercial Aviation in the Jet Era and the Systems that Make it Possible*, https://doi.org/10.1007/978-3-030-20111-1_3

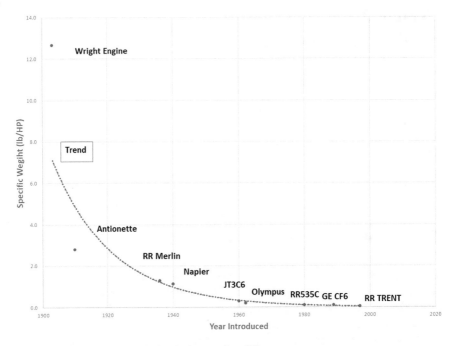

Fig. 3.1 Weight-to-power ratio for airplane engines [1]

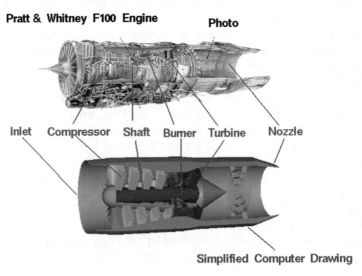

Fig. 3.2 Turbojet engine regions

Fig. 3.3 Transport aircraft
cruise speed progress [2]

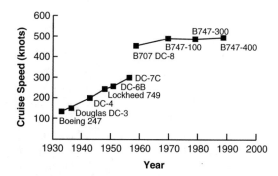

piston-propeller aircraft (introduced 1953). It topped out at about 300 knots airspeed. The Boeing 707 cruised at >500 knots and is an example of a commercially successful jet engine model that began its first flight in 1957. Introducing jet engines meant that in a span of 4 years, the aerospace industry was able to achieve a 60% increase in cruise speed, an enormous improvement with lots of benefits but some drawbacks. This large increase in speed meant a much shorter transit time for passengers, especially on long-range transoceanic or transcontinental flights. The convenience to passengers is obvious from these higher speeds, but airlines soon learned that these higher speeds translated into more revenue flights, as aircraft could now achieve more flights per day, while traveling at the higher speeds from jet propulsion.

The turbojet engine began its flight career in the latter stages of World War II. Both Germany and Great Britain fielded turbojet-powered aircraft, with the German ME262, a twin-engine fighter, being the most iconic and highest performing aircraft of that era. The ME 262 was powered by two Jumo 004B engines, which relied on eight axial flow compressor stages, a ring of six annular combustion chambers and a single axial flow turbine. Each engine could produce nearly 2000 lb$_f$ of thrust [3]. Great Britain countered with the Gloster Meteor. The Meteor's jet engine used a centrifugal compressor, which added complexity to the gas flow path, along with conventional straight wings (vs. the swept wing design of the ME262). The Meteor's jet engine would eventually be produced by Rolls-Royce as the Derwent model. The ME 262 introduced swept wings, axial flow compressors, and underwing pods for engine placement. All three of these design configurations would be duplicated in the commercial jet fleets of the 1960s and beyond. The ME262 did not pioneer tricycle landing gear but did demonstrate its importance to the operation of jet engine aircraft.

While design intricacies would enhance the Jumo engine and its ME262 aircraft vs. its sole WWII competitor the Meteor, its engine life span of about 10 h was woefully inadequate for requirements of war time operation and were inadequate in comparison to the Derwent engine's life of 25 h. This lack of robustness in its design would be a significant detriment to its operational utility, with many ME262s destroyed while on the ground (Fig. 3.4).

The thermodynamic name of the jet engine is the Brayton Cycle. The Brayton model predicts, and operating engines have demonstrated, that an increase in

Fig. 3.4 Jumo 004B powering ME262

efficiency can be achieved by higher compression ratios (compressor discharge pressure/inlet pressure) and higher turbine inlet temperatures. Higher compression ratios require multiple compressor stages and higher strength engine cases. Higher turbine inlet temperatures require new materials and cooling design features in order for the turbine blades to survive the high-temperature gas flow while seeing significant stress from the large centrifugal load inherent in their rotating operation.

While the turbojet engine began its life as a military device, it quickly transitioned to the commercial arena because of its ability to provide high thrust for low weight. It did have trouble spots, some of which remain today. The initial turbojet design in which all the airflow passes through the compressor, combustor, and turbine has a much higher fuel burn vs. piston-propeller and turboprop aircraft. The much higher speed found in jet aircraft quickly outweighed the negative impact of higher fuel burn; however, engine designers are working on new designs to improve the thrust-specific fuel consumption (TSFC, fuel economy) of the present generation of engines. From fundamental thermodynamics, the efficiency of the turbofan engine (that type found on most commercial jet aircraft) can be increased by operating the fan at a lower speed vs. the compressor and turbine section. A new design incorporates a gear box into the engine to lower the fan speed and achieve lower TSFC.

Engine noise greatly increased with the advent of the turbojet engine, and special acoustic surfaces are added to the engine (inlet and exhaust) to quiet the perceived noise both on the ground and in the cabin. Unfortunate for the passengers, the new twin-engine aircraft operating today cannot place the engines further away from the fuselage to quiet the passenger compartment. This is due to the large thrust loads generated by these engines, and the high yaw load that would be placed on the aircraft during one-engine-out operation. Placing the engines further out on the wing would require an even larger tail and rudder surface to overcome this yaw load, in the event of an engine failure during takeoff, landing, or cruise.

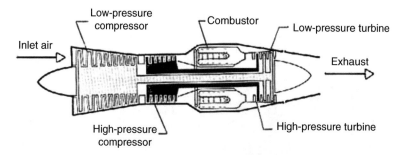

Fig. 3.5 Twin-spool turbojet initially operated on J57 engine for B-52 [4]

Pratt & Whitney is credited with incorporating the two-spool turbojet as part of their design for the then new engine for the Boeing B-52 bomber. The twin-spool design used two axial shafts turning at different speeds inside the engine to provide a more optimal aerodynamic airflow through the various engine sections. In effect it allowed the low-pressure compressor and low (exhaust stages) turbine to run at a lower speed vs. the high-pressure compressor and turbine stages. The two-shaft design added complexity but greatly improved TSFC and overall engine performance. Figure 3.5 shows how these different engine sections can be connected by two shafts operating at different rotational speeds. The implementation requires a hollow shaft (outer) connecting the high-pressure components (Compressor and Turbine) while an inner shaft is rotating inside the hollow shaft to connect the low-pressure components (inlet stages from the compressor and outlet stages from the turbine).

While P&W was developing the two-spool design to increase thrust and specific thrust (Thrust divided by engine weight), GE instituted variable stator vanes in front of the first six (of 17) compressor stages of their J79 engine. The engine had an overall compression ratio of 12:1, like the pressure ratio of the P&W. This variable stator design had the effect of providing improved airflow into the compressor stages at the widely varying flow conditions seen during takeoff, cruise, and other flight scenarios [5].

While these two technical improvements were instituted on military engines, their benefit was equally effective on commercial engines. The two-spool design by P&W evolved into the JT8D-9 which powered the McDonnel-Douglas DC-9 and the Boeing 727 plus 737. The J79 military engine that initiated variable stator compressor vanes by GE was produced in great numbers (>10,000 engines) and allowed them to incorporate this technology into the next phase of gas turbine innovation, the turbofan engine.

It seems that both major US gas turbine propulsion design companies used their DOD contracts to eventually enhance their commercial offerings. GE developed the TF39 engine as parts of its winning bid to power the ultrahigh lift military transport, Lockheed's C-5. This engine introduced the concept of a turbofan, vs. the turbojet seen during the early jet engine days. The turbofan uses a fan (typically in front of the compressor section) to move an even larger volume of air than the volume that

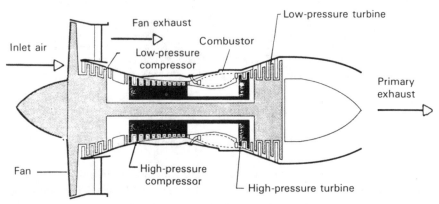

Fig. 3.6 Generic two-spool turbofan engine

passes through the engine. GE's TF39 engine had an 8:1 bypass ratio and a 22:1 pressure ratio. It also adopted the two-spool approach like P&W used for the J57 and continued to use the two-spool design for all future large commercial turbofan engines. Figure 3.6 shows a schematic of a turbofan. Figure 3.6 demonstrates the decrease in fuel consumption that arose from transitioning from turbojets to turbofans in the 1960s.

The 8:1 bypass ratio for the TF39 meant that 8lbs of air would flow through the fan stage and around the engine core (compressor, combustor, and turbine) for every 1 lb. of air that was moved through the core. This engine operated with over 1500 lb./s of total air flowing through both the outer fan duct and engine core during takeoff. It would be shortly after this engine began powering the C-5 that all three major engine companies (GE, Rolls-Royce, and P&W) started offering commercial high-bypass engines.

Rolls-Royce started with the RB-211 which powered several Boeing products including the 747. This engine had a 4.1:1 bypass ratio and a 34.5 compression ratio [6]. P&W offered its JT9D engine with a 4.8:1 bypass ratio and a 26.7 compression ratio [7]. This engine was used on the Boeing 747, 767 several Airbus platforms and the DC-10. Finally, GE produced its commercial variant of the TF39, the CF6-6D in 1971 with a 6.2:1 bypass ratio and a 26:1 compression ratio [8]. While GE instituted variable stator blade positioning and P&W introduced the twin-spool design as their technical response, both companies eventually introduced their competitor's concept into their engine designs. Rolls-Royce introduced the three-spool design (three co-annular shafts) with the RB211. They continue to make a three-spool engine, believing that three shaft rotation speeds will provide a more optimal aerodynamic setup [9]. Both the two-spool and three-spool design introduced added complexity and weight. These additional components required a careful matching of the shaft design, shaft interaction, bearing location, support design, and lubrication.

State-of-the-Art Subsonic Engine SFC

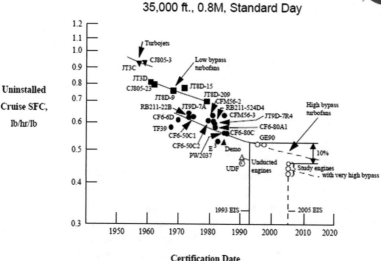

Fig. 3.7 Improvement in TSFC during jet era

Using these moderate bypass ratios, engines were achieving a 35% decrease in fuel burn vs. the original turbojet designs of the 1950s. This marked an important discriminator for airlines as the oil price hikes of the 1970s were making jet fuel costs an even larger fraction of overall ticket prices (Fig. 3.7).

Several improvements were introduced to the jet engine to increase flow stability, reduce fuel consumption, or improve engine reliability. These technologies were the introduction of the multispool approach (two or three rotating shafts) to allow each section of the engine to turn at a better aerodynamic efficiency point. The use of variable stator positions at the inlet of many compressor stages improved airflow and compressor performance across the entire flight envelope. Finally, the switch from a pure jet design (all air flow passes through the engine core) to a bypass turbofan design provided a marked reduction in fuel burn and jet engine noise.

Additional improvements in engine performance after the introduction of the turbofan design in the 1970s included methods to run at higher turbine inlet gas temperatures and higher overall compression ratios. One offshoot of the turbofan has been the need to harden the engine case to prevent a fan blade-out. With the much larger rotating surfaces now at the front of the engine (fan blades vs. compressor blades), the potential for one of these rotating aerodynamic surfaces to be damaged by a bird strike or other foreign object damage (FOD) is high. Therefore, engine designers need to design the engine case to retain the loss of a fan blade. A common certification test is to explosively release a fan blade near its root and verify the containment section holds the fan parts within the engine.

Engineers in the 1970s and 1980s introduced new metal alloys, new methods to form the blades (single crystal growth), and intricate cooling arrangements to allow the turbine section to run at higher temperatures. These improvements have increased the TSFC for the engines of that era. Even greater improvements are expected via a continued increase in turbine inlet temperatures, compression ratios, and higher bypass flow rates. The latest Rolls-Royce model XWB is expected to have a 9.3 bypass ratio. The GE9X will have a 10:1 bypass ratio, combined with a 134 inch diameter fan, it is expected to be the largest turbofan ever made [10].

GE has been investing in ceramic components for their engine hot section, achieving even higher thermodynamic efficiency by the higher operating temperature capability of these materials. P&W has introduced the geared turbofan (GTF), a turbofan engine with a gearbox allowing the fan section to run slower and more efficiently than the compressor stages. Analysts anticipate a 15% reduction in TSFC for the new GTF engine. While GE's offering doesn't seem to have the same level of TSFC improvement, it is based on a less transformational change in engine geometry. Thermodynamics indicates that higher turbine temperatures and compression ratios will increase engine efficiency. However higher bypass ratios will eventually raise to a maximum efficiency point. As bypass ratios increase, fan diameters increase. Larger fans present a larger cross section to flow with larger surface area nacelles. Eventually these increases in fan diameter and nacelle surface area will produce a high enough growth in drag that the increase in bypass ratio will actually reduce TSFC.

While P&W has been the first to introduce the GTF to produce a reduction in TSFC, other engine companies are expected to produce their own variant of the GTF. One consequence of the transition to GTF engines is the increase in fan diameter and bypass ratios. The larger fan diameter has a consequence on other airplane design features. For incorporating GTFs in existing airframe designs (e.g., A320 neo, new engine option) landing gear as well as engine pylon and nacelle interact. Larger fans translate to larger nacelle diameters, which may not have sufficient ground clearance with existing landing gear design. Future nacelles are envisioned with shorter pylons to accommodate these larger fan diameters without requiring extended length landing gear. Shorter pylons will require new analyses of the engine/nacelle combination operating in closer proximity to the wing. New methods of insuring sufficient lift from the wing with a nacelle sitting very close to the lower surface will be required in future designs [11].

ETOPS

The FAA (Federal Aviation Administration) and EASA (European Aviation Safety Agency) institute rigorous aircraft maintenance and flight operating rules for safe travel. Their precursor groups along with the aviation industry have had a history of trying to improve aircraft flight safety. Early commercial flights operated with the principal that they would never be over 100 miles from an airport that they could divert to in case of an in-flight emergency.

As commercial jet engines increased in thrust and reliability, the FAA instituted a program called ETOPS, which originally meant Extended Twin-engine OPerationS. This rule, as originated, stated that a twin-engine commercial aircraft could never be more than 1 h from an appropriate airfield. The 1 h flight time was under the reduced velocity of single-engine operation. The rule meant to protect passengers in case of one engine failure, the airplane could safely reach a diversion airport.

As engine reliability increased, the agency has allowed operators to extend this time and distance. Certain aircraft/engine types are now certified to 240 min ETOPS, or 4 h flying distance. The airlines prefer longer ETOPS time, as it allows for more direct flight paths, with much lower fuel costs. While originally targeted for two-engine aircraft, ETOPS has now been extended to three- and four-engine aircraft. As most of the ETOPS flight rules apply to long transoceanic flights, industry personnel began to refer to ETOPS as "Engines Turn Or People Swim."

The early jet transport aircraft (De Havilland Comet, Boeing 707 and Douglas DC-8) relied on four wing-mounted engines. In the USA, Boeing and Douglas used pylons to support the engines beneath the wing, while De Havilland used a more aerodynamic location within the wing root. While the pylon-mounted design has continued to be used, the De Havilland design with the engine embedded in the wing provided less aerodynamic drag vs. the underwing hung alternative of the Boeing and Douglas aircraft. The major negative feature of the De Havilland design arose from the much higher maintenance required by jet engines in the 1940s and the difficulty in reaching those engines while embedded within the wing [12]. A pod-mounted engine slung below the wing provided better and more rapid maintenance vs. the De Havilland design.

One method to track the transition from the early four-engine designs, to the three-engine designs of the 1960s and today's two-engine designs is through the increase in thrust available from these engines through time. The early transports relied on four engines because of the low thrust level and reliability in these engines. A four-engine design was required for the Boeing 747 because of the high load and drag inherent in this ultra-large aircraft. However, three engines were found in their slightly smaller wide-body brethren, the McDonnel Douglas DC-10, and the Lockheed L-1011. Three engines were also employed for smaller aircraft like the Boeing 727. An eight-engine configuration must contend with the difficulty in placing an odd number of engines. A two-engine, four-engine, or even eight-engine design (like the B-52 bomber) can place the engines in underwing pods with an even distribution about both sides of the aircraft centerline. Some aircraft designers have opted to place the engines near the tail, where the aircraft remains easier to control in the event of a loss of engine function. The yawing forces are much lower when the engine thrust is so close to the aircraft centerline. However, the third engine is commonly placed on the aircraft centerline, with the DC-10 opting for a straight through design and the 727 using an S-shaped duct to place the engine thrust in-line with the fuselage. The S-shaped duct makes for higher aerodynamic losses but limits the downward force from this center-placed engine.

Today most aircraft rely on two engines, due to the higher thrust levels that can be achieved in these high-bypass-ratio engines. In addition, engine reliability has increased significantly allowing operators to fly these aircraft for longer periods of

Table 3.1 Growth in engine thrust

Engine Mfr/model	Type	Year	Thrust (Lbf)
RR/Conway	Turbojet	1960	20,000
GE/CJ805	Turbofan	1955	16,000
PW/JT3D	Turbofan	1960	17,000
RR/RB211	Turbofan	1969	51,000
GE/CFM56	Turbofan	1974	20,000
PW/JT9D	Turbofan	1966	45,600
RR/Trent800	Turbofan	1995	90,000
GE/90	Turbofan	1993	81,000
PW/4000	Turbofan	1986	50,000

time without in-flight shutdowns. Table 3.1 shows the general increase in commercial engine thrust through the decades from the three major engine suppliers (GE, P&W, and Rolls-Royce).

Engine Operating Envelope

The turbofan engines designed by the three large engine companies (GE, Rolls-Royce, and P&W) have been designed and tested to operate in a wide range of ambient conditions. All these devices rely on a large volume and a large mass of air entering the engine to achieve their rated thrust. However, on extremely hot days the inlet mass flow rate will be lower, because of the drop in air density with increasing temperature. This mass flow drop, coupled with the lower lift endemic with lower air density, can be troublesome for certain airplane/engine combinations. The CRJ, a smaller regional jet produced by Brazilian Embraer, has only been certified to 118 F. Tens of flights were cancelled at Phoenix's Sky Harbor airport in June 2017 when the ambient temperature climbed to 120 F [13].

Not surprisingly, as the main recipient of the aircraft fuel system and the only desired location with flames, engineers incorporate many safety systems into the engine and nacelle. The nacelle acts to limit fire spread by incorporating design features to prevent fire from spreading outside the nacelle. The engines have flame detectors outside their combustion zones. An engine fire indication will be met by closing the fuel supply to the subject engine and the immediate application of fire extinguisher material (Halon or other flame smothering gas).

Nacelles

The nacelle is a housing that surrounds the gas turbine engine. It is interesting to note that the ME262, the first operational jet aircraft built in significant numbers, also introduced the concept of the nacelle surrounding the engine. The ME262 also started the now common practice of installing the engines and nacelles on pylons hanging below the main wing. The commercial airline industry has now evolved such that almost all aircraft now rely on two engines slung under a sweptback wing in a nacelle, the identical design used by Germany for their ME262 fighter almost 75 years ago.

The nacelle has several functions during flight and throughout the engine operating envelope. The nacelle surrounds the engine, protecting it from the elements (hail, lightning, etc.). The nacelle also provides most of the noise attenuation for the engine. The nacelle will provide fire containment and suppression in the unlikely (now) event of an engine fire. Most important for propulsion, the nacelle provides the bypass duct for the fan flow that now provides most of the engine thrust. This bypass fan duct will see >85% of the total airflow and provide a similarly high fraction of the engine thrust. Finally, the nacelle will generally incorporate a thrust-reversing mechanism to aid in aircraft stopping.

The nacelle generally protrudes in front of the fan (the first rotating component in a turbofan engine). The space between the nacelle leading edge and the fan is meant to allow for an even distribution of air throughout all quadrants of the engine face. The surface of the nacelle is designed to be smooth to minimize aircraft drag. The popular configuration of placing the engines in nacelles below the wings make the engines and nacelles the second largest drag producing (check Fact) item on the aircraft (Fig. 3.8).

The nacelle liner (the surface of the fan duct inboard and outboard annulus) plus the inlet has been adopted as a convenient location for noise abatement. The early jets which exhausted high-temperature gas at a very high velocity (thus generating thrust for propulsion) also generated significant noise from this high-energy exit gas. The solution has been to transition to the turbofan engine which now has a

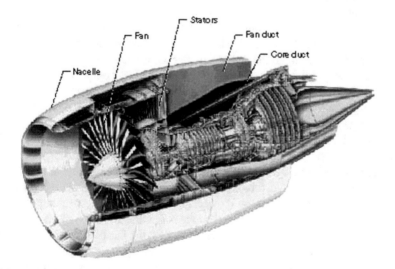

Fig. 3.8 Engine face showing relative thickness of nacelle to improve air flow

slower moving, cooler gas stream surrounding the hot gas leaving the engine core. This outer gas both mitigates the noise from the entire engine, but even turbofan engines rely on noise abatement surfaces within the nacelle to reduce engine noise even further. The noise abatement comes from small holes in the surface of the nacelle. These holes are sized and tuned to the engine noise frequency. Using a process called Helmholtz resonance, they cancel specific noise regimes. These surfaces add drag and complexity to the nacelle but provide significant noise abatement from the gas turbine.

All the aircraft surfaces need to be protected from lightning strike. The nacelles together with the cockpit are the regions of highest likelihood for lightning attachment. Therefore, they must be afforded the highest level of protection. A common method for lightning protection uses a highly conductive mesh (e.g., copper) near the surface of these likely lightning strike zones. This conductive mesh is tied into a ground loop which transfers the lightning energy through the aircraft, allowing it to reenter the atmosphere at an alternate position on the aircraft. This technique was easier when aircraft were predominantly aluminum, a good electrical conductor. The entire vehicle acted as a Faraday cage allowing the electrical energy to transfer through the skin of the aircraft and reenter the atmosphere at some alternate location. Newer predominantly CFRP aircraft need to institute the conductive mesh listed above. This feature is important as the average commercial aircraft gets struck by lightning approximately once per year [14].

Other important nacelle functions include allowing access to the engine for maintenance and inspection. This requirement has led to most engine installations being underwing. Finally, the nacelle can be reconfigured through an actuator, to achieve reverse thrust. This technique is especially useful on wet or icy surfaces when the runway-tire interface has the highest incidence of skidding.

The first function of the nacelle is to house and protect the engine, while also providing efficient air flow passages for the inlet and exit streams. For the turbofan engines found in the vast majority of today's jet airliners, there is a single inlet stream leading to the fan section. The inlet of the nacelle duct needs to provide a uniform flow stream to the fan so that the fan section does not become overloaded or starved in any quadrant. The flow uniformity can be complicated especially at takeoff and landing by fixed runway directions with significant crosswinds. The design response is a thick-walled, lengthy inlet duct that allows the flow to become uniform regardless of crosswinds and aircraft speed.

An important feature of the nacelle is its ability to incorporate a thrust reverser into the engine operation (after touchdown, see Chap. 10). The thrust-reversing mode allows the significant engine thrust developed in a high-bypass-ratio turbofan engine to be primarily redirected forward, thus aiding in stopping the aircraft. This is the reason that you will hear the engines spool up (power up) from the low power setting of descent. With most of the thrust developed in the bypass fan flow in these turbofan engines, a thrust reverser can be employed that only redirects the fan flow forward. The thrust reverser does not alter the core engine flow, the air that is going through the compressor, combustor, and turbine section. However, it is still effective as thrust-reversing mechanism, with the engine core flow still allowed to maintain its rearward travel [15].

Early thrust-reversing mechanisms employed a clamshell device at the end of the nacelle. This device rotated two halves into position redirecting the flow from aft toward the front of the aircraft. Newer thrust reversers employ a sliding sleeve to both block fan duct flow and open louvers to redirect the flow forward. The newer designs allow a more precise direction for airflow, avoiding flow down to the runway or onto the fuselage. Military transports have made novel use of the thrust-reversing feature. The C-17 can safely operate their thrust-reversing (TR) system in-flight in order to achieve a steeper glide slope and more rapid landing. The same TR system on the C-17 can move the aircraft backward (on the ground), something not designed in commercial transport aircraft [16]. Some other military TR systems can be employed in-flight to enhance the maneuverability of fighter aircraft. Figure 3.9 shows the cascade-type thrust-reversing scheme which has become the dominant style among turbofan-powered commercial aircraft.

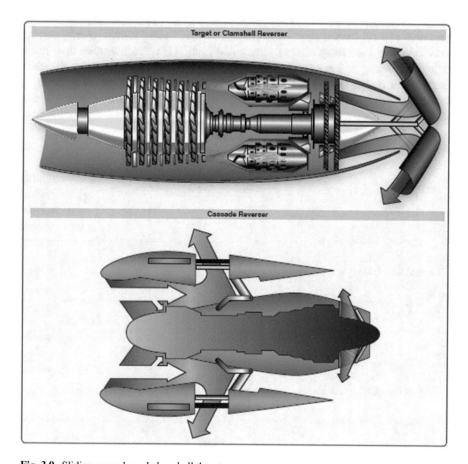

Fig. 3.9 Sliding cascade and clamshell thrust reverser

It is interesting to note that the FAA, EASA, nor any other aviation regulatory body requires thrust reversers on planes. A similar feature was incorporated into propeller craft when variable pitch propellers increased propeller efficiency. Varying the propeller pitch (angle or bite) of the blades at various power settings allowed them to be more efficient aerodynamically. This feature also allowed them to be rotated to generate negative thrust upon landing, the first thrust-reversing system. While not mandated by any aviation supervisory agency thrust reversers provide a strong margin of safety when landing on below optimal runways (wet, snowy, or icy). In addition, their work can save brake life by transferring some of the stopping energy into the engine, versus shunting it all into the brake materials.

Both landing speeds and airplane weights have increased in the transition from propeller craft and have continued to grow in the turbojet to turbofan era. Figure 3.10 below shows the increase in the kinetic energy required to be absorbed during the landing of these vehicles during the early days of the jet era. Converting all of this kinetic energy into brake heating may be possible (and is required to achieve flight certification) but will rapidly wear out the earlier metal brake designs, and still challenges the carbon-carbon material used in most large passenger planes today. Thrust reversers can off-load a significant percent of the landing kinetic energy, saving brake wear and reducing aircraft turnaround time. Aircraft turnaround time (time between landing and ready for takeoff) can sometimes be limited by the time required for brakes to cool to a minimum temperature before taking off following a landing. Brakes must always be available to handle a rejected takeoff. The massive

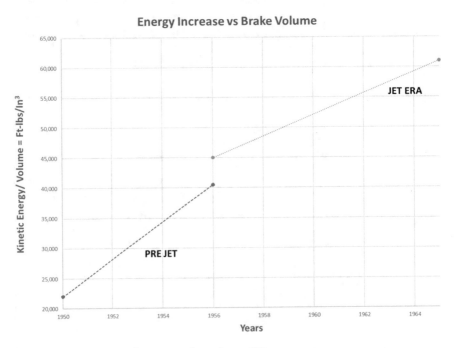

Fig. 3.10 Landing energy from pre- and post-jet era [17]

amount of energy required to stop a plane from a rejected takeoff can only be handled if the brakes do not retain too much heat from the previous landing. Thrust reversing can direct some of the landing energy away from the brakes keeping them at a lower temperature and allowing them to cool to their maximum takeoff temperature faster. For rejected takeoff analysis and design, the effects of the engine thrust reversers are not factored into the event.

The engine nacelles like the wings can suffer aerodynamic losses if ice accumulates on their surface. In addition, ice accumulation on the inner diameter upstream of the fan could damage or cause fan failure if ice built up then shed into the fan blade. For this reason, the first few inches of the nacelle inner barrel are considered a deice region. One convenient way to achieve this deice is to direct air from one of the compressor stages of the engine into the inside of the nacelle leading edge. The air has been heated because of the compressor work and circulating this hot air inside the nacelle leading edge will prevent ice from forming. For instance, air from the seventh compressor stage of the P&WF100-PW-220 turbofan engine (used on the F-15 and F-16) will be about 450 F [18], hot enough to prevent ice formation, and if not directed properly, hot enough to soften the aluminum inlet of the nacelle. Hence, the method of circulating this bleed air around the leading edge of the nacelle is important to retain its structural integrity.

While to the flying public the engines and nacelles may appear as one entity, they are different features, made by different companies and then integrated at the airframe facility (e.g., Seattle, Boeing, Toulouse, Airbus). General Electric, Pratt & Whitney, and Rolls-Royce are the three western companies remaining in the large turbofan market. Aircelle, Safran, and UTAS Aerostructure division are some of the nacelle suppliers to this same large turbofan market.

References

1. Advances in Aircraft Engines since 1903, Scivaraj, R., Kodanda, B., Proceedings of 100 Years since 1st Powered flight Seminar, Bangalore 2003
2. IPCC Report Aviation and the Global Atmosphere, 1999, Figure 7.1 p 435.
3. https://airandspace.si.edu/collection-objects/junkers-jumo-004-b-turbojet-engine retrieved Aug 19, 2017.
4. Dependable Engines, the story of Pratt & Whitney, Sullivan, M. AIAA, 2008, ISBN 978–1–56347-957-1
5. The History of the Aircraft Gas Turbine Engine Development in the United States … A Tradition of Excellence, St. Peter, J. ASME, 1999, ISBN 0–7918–0097-0
6. https://www.rolls-royce.com/products-and-services/civil-aerospace/airlines/rb211-535e4.aspx#downloads retrieved 8/22/2017
7. http://www.pw.utc.com/JT9D_Engine retrieved 8/22/2017
8. Turbine Engines of the World, Flight International, January 2, 1975
9. GAS TURBINE TECHNOLOGY EVOLUTION - A DESIGNER'S PERSPECTIVE, Koff, B., AIAA 2003–2722
10. Aviation Week & Space Technology, May 9–22, 2016, P. 73

11. Clean Sky 2, 4th Call for Proposals, Development of scaled models for Synthetic Jet Actuators based on Aerodynamic Characterization in CFD, Ground and Wind Tunnel Testing, JTI-CS2–2016-CFP04-LPA-01-27, May 19, 2016
12. Embedded Wing Propulsion Conceptual Study, NASA TM 2003–212696, Kim, H., Saunders, J.
13. https://www.forbes.com/sites/marshallshepherd/2017/06/20/the-science-of-why-its-too-hot-for-some-planes-to-fly-in-the-southwest-u-s/#7a10d62554ce retrieved 9/6/17
14. https://www.scientificamerican.com/article/what-happens-when-lightni/ retrieved 8/13/18
15. Aircraft Propulsion 2nd Ed., Farokhi, S., John Wiley & Sons, 2014, ISBN 978–1–118-80677
16. http://www.airforce-technology.com/projects/c17/, retrieved Sep 10, 2017
17. New Designs for Commercial Aircraft Wheels and Brakes, Stanton, G., Journal of Aircraft, Vol. 5, No. 1 Jan-Feb 1968
18. The Effects of Compressor Seventh-Stage Bleed Air Extraction on Performance of the F100-PW-220 Afterburning Turbofan engine, Evans, A., NASA Contractor Report 179447, Feb 1991

Chapter 4
Cabin Pressurization and Air-Conditioning

The history of aviation, both military and commercial, documents numerous records that focused on increasing altitude and top speed, especially during the first three decades of flight before WWII. The military advantage of these two attributes seems obvious, but both increasing airspeed and the ability to fly higher imbued benefits to commercial aviation as well. The convenience to passengers of faster travel seemed obvious, but the increased speed also meant that aircraft were now available to complete more connections, increasing the revenue available from the faster airliners.

The impetus for higher altitudes originally came from safety concerns and the desire to fly over inclement weather. In the USA, the Rocky Mountain range provides a significant barrier to West-East travel with peaks easily reaching 15,000 ft. As the cruising altitude and especially the cruising speed of airliners increased, an increase in actual flight altitude would generally translate to lower fuel burn. The lower air density at higher altitude generated less drag on the aircraft, thereby reducing the amount of fuel required. With the arrival of jet engines, their much higher propulsive force vs. propellers (thrust), allowed these new airframes to reach higher speeds, but they also gobbled fuel at a much higher rate. This natural fuel burn penalty associated with the propulsion system was magnified because the drag force on an object (airliner) is a function of the square of its velocity. A nearly 50% increase in average speed 325 mph (propeller era) vs. 500 mph for the jet era produced over a 2× increase (drag ~ velocity squared, $1.5^2 = 2.25$) in the airplane drag plus the added thrust required to move the platform at this higher velocity. However, the drag force is also proportional to the air density. Higher altitudes mean lower density, less drag, and lower fuel burn [1]. So while early passenger planes sought higher altitudes to avoid weather, flying at over 30,000 ft. provides a significant fuel saving for jet-powered aircraft.

Figure 4.1 below shows the average air temperature at various altitudes within the troposphere. The troposphere is the bottom layer of our atmosphere, where the temperature drops in a nearly linear fashion as you increase in altitude. The troposphere, the tropopause, and the lower region of the stratosphere is where all commercial air traffic occurs. Not shown in Fig. 4.1 is the tropopause and the stratosphere.

© Springer Nature Switzerland AG 2020   45
T. Filburn, *Commercial Aviation in the Jet Era and the Systems that Make it
Possible*, https://doi.org/10.1007/978-3-030-20111-1_4

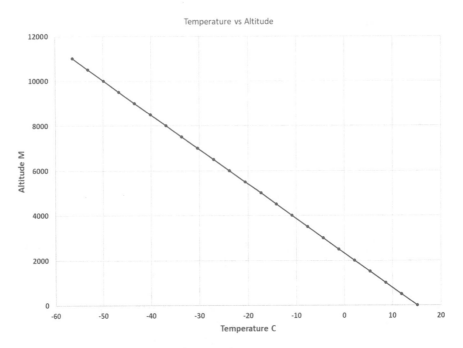

Fig. 4.1 Air temperature vs. altitude for troposphere

The tropopause is the region between the troposphere and the stratosphere, while the stratosphere is the region above the troposphere, reaching an altitude of about 30 miles (>150,000 ft.) well above the altitude of commercial air traffic. The thin tropopause region retains a relatively constant air temperature. Since the retirement of the Concorde, the stratosphere only sees commercial airline traffic in its lowest regions, but it also has a nearly constant air temperature in its lower regions, where air traffic flies. It does see a linear increase in temperature at its upper levels, until it reaches the mesosphere. The average temperature for a plane flying at 41,000 ft. (one of the higher commercial altitudes except for Concorde, but lower regions of the Stratosphere) is about -70 F.

So it is obvious that a pilot in an open cockpit would need protection from the very cold ambient temperatures above a height of ~ 3000 ft. (~1 km) where the temperature is at 10 C (50 F) or lower. The effect of this cold is exacerbated by the additional cooling effect of the high-velocity air circulating in an open cockpit. By World War I, and certainly during the Air Mail heyday after the war, pilots learned to dress for the thermal challenge of their unpressurized and open vantage point.

As flight times and aircraft altitudes increased, it quickly became apparent that simply adding layers to the flight crews clothing was not adequate to maintain pilots warm enough to perform their duties. Airplane designers sought a method to provide for the thermal comfort of the flight crew and passengers, with the high-temperature engine exhaust seeming to be a reasonable source for adding heat to the cabin.

Table 4.1 Time of useful consciousness vs. altitude

Altitude (ft)	Atmospheric pressure (psi)	Atmospheric pressure (mmHg)	Oxygen partial-pressure (mmHg)	Temperature (°F)	Time of useful consciousness
100,000	0.15	8	2	−51	0
90,000	0.25	13	3	−56	0
75,000	0.50	27	6	−65	0
63,000	0.73	47	10	−67	0
50,000	1.69	88	18	−67	0–5 s
43,000	2.40	123	26	−67	5–10 s
40,000	2.72	141	30	−67	10–20 s
35,000	3.50	179	38	−66	30–60 s
30,000	4.36	226	47	−48	1–3 min
25,000	5.45	282	59	−30	3–5 min
20,000	6.75	349	73	−12	10–20 min
18,000	7.34	380	80	−5	20–30 min
15,000	8.30	429	90	5	30+ min
10,000	10.11	553	116	23	Nearly indefinitely
7000	11.30	587	123	34	Indefinitely
Sea level	14.69	760	160	59	Indefinitely

Data for an ISO standard day (59 °F/15 °C) at 40° latitude
Source: U.S. Naval Flight Surgeon's Manual, Naval Aerospace Medical Institute, Third Edition, 1991

After World War I, the use of the open cockpit started to fade. At this time airplanes routinely reached altitudes that forced pilots to limit exposure to the low air pressure and most important for pilot consciousness, low oxygen pressure. When aircraft started flying above 20,000 ft. altitude, the time of useful consciousness was found to be under 10 min without countermeasures (e.g., supplemental O_2 or a pressurized cabin). Table 4.1 shows the time of useful consciousness vs. altitude as determined by the Navy's Flight Surgeon Manual [2], pilots in the 1920s did not have as detailed a chart as Table 4.1 but understood the troubles of higher altitude flight. Mountaineering efforts, balloon flights (including 1875 balloon flight that killed two of three occupants after reaching 28,000 ft) confirmed pilot and passenger anecdotes about the difficulties in reaching higher altitudes. In 1935 the US Army Air Corps prepared a Technical Report "The Physiological Requirements of Sealed High Altitude Aircraft Compartments." This report recommended three potential solutions to the problem:

1. Pressurized compartment with air.
2. Pressurized compartment with oxygen.
3. For altitudes between 15,000 and 40,000 ft., an unpressurized compartment with only oxygen.

Researchers and aircraft designers recognized the flammability dangers of oxygen only compartments (especially for military aircraft) and thus focused on developing pressurized compartments with air [3].

The FAA and other aviation regulatory bodies recognize the extremely limited time available for pilots to respond in the event of a depressurization accident at the typical cruising altitudes found in modern-day airliners (>30,000 ft.). For a common setup with a pilot and first officer, it requires at least one person in the cockpit to have a mask on when flying at 35,000 ft. or higher. This requirement becomes a mandate to don a mask at 25,000 ft. if one of the pilots leaves the cockpit for any reason [4].

Our atmosphere contains approximately 21% oxygen by volume. This percentage does not change with altitude, but as the altitude increases and the total pressure decreases the amount of oxygen available (O_2 partial pressure) drops. Figure 4.2 shows this change in O_2 partial pressure and how it creates hypoxic conditions for pilots or passengers at altitudes above ~12,000 ft. A horizontal trace at whatever altitude you choose will identify whether it is healthy or hypoxic or hyperoxic (too much oxygen). In this case 12,000 ft. and above is the hypoxic region for 21% O_2 concentration (earth's atmosphere). Hypoxia is the condition where insufficient oxygen is in your bloodstream. There is a third region of Fig. 4.2, the hyperoxic region, where too much oxygen can also damage tissues and potentially lead to death. It is the concern over hyperoxia that the emergency oxygen supply should be discontinued when the aircraft has descended below 15,000 ft. This would

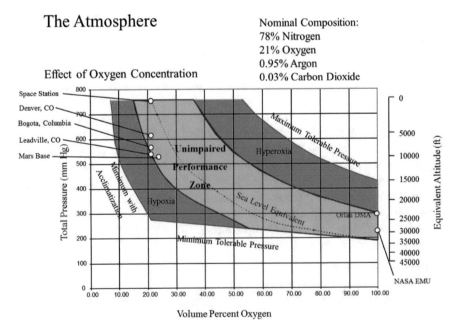

Fig. 4.2 Atmosphere and regions of impaired and unimpaired performance

correspond to the 100% oxygen condition and is demonstrably in the hyperoxic region at 15,000 ft. and lower altitude. Figure 4.2 is somewhat nonintuitive in that lower altitudes are represented by higher points on the Y-axis.

As the hypoxia problem became evident, the solution seemed to swing between the cost and weight of pressurizing the entire aircraft or providing a pressure suit for the crew. Wiley Post made newspaper headlines while producing and flying in the first operable pressure suit for a pilot [5]. His pressure suit let him breathe at an equivalent altitude 5000–10,000 ft. lower than his actual altitude. This novel suit allowed him to remain conscious while operating at near record altitudes for long periods. His suit did restrict his mobility, a problem that would plague pressure suit designers, including NASA up to the Apollo program and beyond. Wiley Post's suit did allow him to demonstrate the "river of air" (term coined by Wiley) moving at high altitudes in our atmosphere. He apparently provided convincing evidence for the jet stream by achieving flight speeds >100 mph over what his Lockheed Vega was capable of at lower altitudes. A small percentage of that increase came from reduced friction, but the vast majority came from the propulsion boost achieved by flying with the jet stream (generally west to east).

Wiley Post did not set any recognized altitude records, but his pioneering work on pressure suits along with his early demonstration of the jet stream would contribute greatly to the aerospace community. It is most unfortunate that his last flight ended in a crash in Alaska with famed American humorist Will Rogers on board.

Boeing provided airlines with the first pressurized cabin for passenger service when it introduced the model 307 Stratoliner. Boeing only produced ten of this type, as its introduction in Dec 1938 was very unfortunate timing, with World War II beginning only 9 months later. The four-engine Boeing airliner could fly as high as 26,200 ft., and cruise at 220 mph. It had a crew of five and could accommodate 33 passengers while traveling nearly 2400 miles [6]. A 26,000 ft. ceiling may not seem high compared to today's jet airliners, but most of the turbulence and resulting passenger discomfort was found below this altitude. In fact, the first flight attendants on commercial airliners were nurses to help passengers through the nausea they experienced due to the frequent turbulence seen in lower flying aircraft.

Those in the aerospace community during the 1930s knew that both people and engines needed a boost in air pressure when flying at high altitudes. Piston engine designers created the supercharger to increase the pressure and mass of air entering the engines at high altitudes. Higher air mass flows meant that additional fuel could be added to each piston, which greatly boosted the engine power available at these altitudes. Figure 4.3 shows a schematic of a generic supercharger at work.

A supercharger is an air pump, geared off the crankshaft (power shaft driving the propeller) to a turbine. The gearing allows the compressor to operate at its preferred rotational speed, which usually differed from the engine rotational speed, which was designed for optimal propeller performance. The system uses air from a Ram air duct to provide a slight boost in pressure for the air supply to the supercharger. Ram air is a specially designed duct for ingesting ambient (low-pressure air from outside the aircraft). The ram air duct is useful for converting the high velocity of entering air, into increased air pressure, reducing the requirement for additional

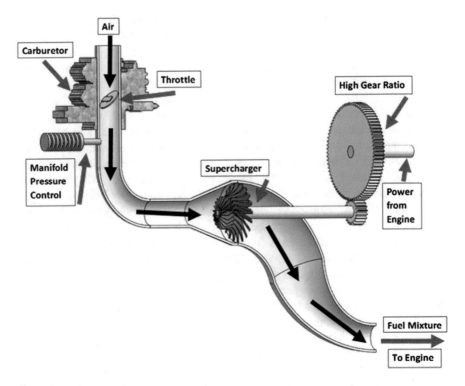

Fig. 4.3 Engine supercharger

compression. Fuel is usually added upstream of the supercharger (compressor) and then a heat exchanger (not shown in Fig. 4.3) cools the air-fuel mixture after it is compressed in the supercharger. This heat exchanger reduces the heat picked up by the air during the compression process. Air follows the Ideal Gas Law, so a lower air temperature will automatically increase the density of the mixture entering the cylinder. Increased air density meant higher fuel flow which equated to higher power, important in civil aviation, but vital for military aircraft.

Boeing engineers used a similar compressor arrangement (minus the fuel addition) to create the first commercial cabin pressurization system for the model 307. Several aircraft design changes had to be incorporated to achieve this cabin pressurization including a neoprene seal around the rivet joints of the fuselage, incorporating round Plexiglas windows (both items helped to sustain the inside/outside pressure differential), and having a gas valve in the tail to provide a continuous outflow of gas. Many of these initial Boeing design solutions have survived through today's aircraft, including an outflow valve in the tail, and round Plexiglas windows.

The Boeing system relied on a variable pressure regulator, so that as the aircraft ascended above 8000 ft., the internal cabin pressure would be maintained at 8000 ft. The pressurization system could not keep this equivalent altitude if the aircraft flew above 16,000 ft., the internal pressure would drop linearly with increasing altitude above this 16,000 ft. mark. When the aircraft reached its service ceiling, 20,000 ft., the pressurization system would keep the cabin at a pressure equal to 12,000 ft., still sufficient for passenger safety.

The model 307 system incorporated a heater to warm the air prior to introduction to the cabin. It is interesting to note that the 8000 ft. equivalent altitude has been retained by most passenger airliners during the jet era. Recently both Boeing and Airbus have begun offering aircraft models that keep a higher pressure, lower equivalent altitude cabin, 6000 ft. altitude. The cabin pressure is always a tradeoff between passenger comfort and the weight penalty that higher pressure, lower altitude cabins entail. Any increase in cabin pressure will require a stronger, and in the past thicker walled aircraft. Thicker walls equal heavier fuselages, and weight is the bane of all aerospace engineers. Newer composite materials have allowed the aircraft designers to increase passenger comfort through new, higher strength, carbon fiber reinforced plastic (CFRP) materials, without increasing aircraft weight. However, the Federal Aviation Regulation (FAR) still requires commercial aircraft to pressurize their cabin to a maximum altitude of 8000 ft. equivalent pressure [7].

Most commercial passenger jet aircraft have limited their service ceiling to about 41,000 ft. with some slightly higher excursions. The supersonic Concorde flew at 51,000 ft. because of the much lower aircraft drag that would be found at the lower air density commensurate with that increased altitude. The Concorde flying at Mach 2 flew over 2× faster vs. conventional subsonic airliner that fly at Mach 0.85–0.9. The reduced drag from the higher Concorde altitude was very important for fuel economy. In addition, the higher drag on the Concorde at lower altitudes would have increased the skin temperature. The Concorde fuselage was designed to handle outer surface metal temperatures of nearly 210 F, these high temperatures were routinely seen at its elevated cruising altitude. Higher metal temperatures would have been seen at lower altitudes due to increased skin friction. These higher skin temperatures would have produced lower material strengths due to the higher temperatures or required Concorde designers to switch to heavier metal alloys that retained their strength at these higher temperatures.

As usual in engineering trade studies, the ~40,000 ft. ceiling for commercial subsonic airliners (Airbus A380 [8] 43,000 ft., Boeing 777 [9], 38,000 ft.) represents a compromise between low-temperature, low-pressure and therefore low-drag air, and the increase in lift needed (larger wing area) to support the airplane at higher altitudes. In addition, the increasing gas temperatures at the higher levels of the stratosphere work to decrease engine efficiency in this region above the tropopause (the boundary region between the stratosphere and the troposphere). As mentioned previously, higher gas temperatures produce lower gas density, thereby lower mass flow rates.

The rate at which the cabin achieves its operating pressure (8000 ft. equivalent altitude, depressurizes) and returns to the outside pressure (repress) is important to passengers. This pressure change seems especially noticeable upon descent (repress). Illness and blocked sinuses can heighten the sensitivity to these pressure changes. Therefore, the cabin pressure rate of change is generally limited to about a

300 ft./min [10]. Many early pilots used this 300 ft./min descent rate as their guide to prevent passenger distress in their unpressurized aircraft. This 300 ft./min is much slower than the initial airplane descent rate which can be five times this rate, but goes unnoticed by passengers as the cabin pressurization system stays constant at the higher altitudes.

Another trade study used in determining the operating altitude of aircraft is the oxygen supply system, and emergency oxygen supply. High flying and highly maneuverable military aircraft use a dedicated oxygen supply for the pilot with a tightly conforming face mask being used to deliver oxygen to the pilot. This mask can continue to supply high-pressure oxygen to the pilot even if the cockpit pressure drops to the same pressure as the outside pressure. For these combat aircraft, their operating envelope includes regions (extremely high altitudes) that could produce a loss of consciousness within seconds. Therefore, pilots in these types of aircraft continuously wear their oxygen masks.

The oxygen masks that can be deployed on commercial aircraft for emergency passenger use differ markedly from the systems worn by combat pilots. These masks do not create a pressure-tight seal and therefore can only achieve oxygen supply pressures equivalent to the local ambient pressure. For commercial aircraft in a rapid depressurization event, this will be the pressure of the air outside the fuselage. Hence, flying above 45,000 ft. can produce such rapid unconsciousness that passengers would not have time to don their masks. In addition, these emergency oxygen candles cannot generate high pressure, so the low-pressure pure O_2 stream would not have enough pressure to keep the passengers alert. The high-pressure conformal masks worn by pilots are deemed too expensive and too complicated to don for the flying public. So, one condition that limits commercial aircraft flight altitude is the need to allow passengers to retain consciousness in the event of a rapid depressurization event (loss of a window is a typical event consideration). The Concorde flew over 1 mile higher than the altitude of subsonic commercial airliners. It was granted a waiver for these normally unsuitable altitudes based on analysis of the amount of time it would take to reach low pressure within the cabin in the event of a window failure. While the Concorde passenger cabin was small with typical seating for 100 passengers, the windows (the most likely pressure boundary failure point) were also small. The pressure decay time for the Concorde was deemed sufficiently long for the pilots to reach a lower altitude sufficient for passengers to retain consciousness and don their conventional, generally yellow, oxygen masks.

The oxygen supply for military pilots, commercial pilots, and commercial passengers comes from different sources. On-board oxygen generating systems have been used on many military platforms (C-17, F-35). These systems use pressure swing adsorption (PSA) systems to concentrate the oxygen from ambient air for pilot use. Conventional high-pressure oxygen cylinders are the predominant method for storing oxygen for commercial pilot use. Chemical oxygen generators (chlorate candles) dominate as the source for emergency passenger oxygen. The different systems arise from the varied requirements. Commercial airliners rarely need their passenger oxygen systems; therefore they are optimized for low weight, but reliable oxygen generation. Military systems have frequent, sometimes lengthy periods of use. Hence, they have found benefit from PSA systems that can derive O_2 from ambient air. The crew on commercial airliners uses oxygen intermittently but need to store sufficient O_2 for any crew emergencies.

The present generation of commercial jet aircraft generally relies on air cycle machines (ACM) for conditioning the air in the cabin. Unlike terrestrial systems which rely on vapor cycle machines (refrigerant that goes from liquid to vapor) for cooling, airplanes have predominantly adopted the ACM as their air-conditioning and cabin pressurization system. While vapor cycle machines generally have higher operating efficiencies, the ACMs can achieve much lighter system weights due to the lower density and mass of refrigerant (air vs. Freon-type liquids), again lower weight provides a significant benefit for the ACM use in aerospace applications.

The ACMs (up until the Boeing 787) take air from the engine compressor as their starting fluid. During normal operation the air is "bled" from a low compressor stage, and if insufficient in pressure, can be taken from a later higher pressure compressor stage. While similar in concept to the supercharger air first used on the Boeing 307 Stratoliner, bleed air has begun to travel through the engine, whereas supercharger air had not traveled into the piston engines of the Stratoliner era. The supercharger was a mechanical compressor designed to compress the air solely for the cabin environment. Bleed air for the ACM is a small fraction of the enormous air flow that cycles through the compressor, then combustor and finally turbine section of the jet engine. As this is engine air, any withdrawal affects the overall engine performance and in fact reduces TSFC by its parasitic removal of compressed air from the engine. Figure 4.4 shows a simple three-wheel device that demonstrates the overall operation of the ACMs. This figure shows the myriad of compressors, turbines, and heat exchangers that are grouped together to create the ACM for cabin comfort and pressurization.

The three-wheel ACM of Fig. 4.4 uses two air streams to achieve cabin pressurization and an appropriate air temperature within the passenger compartment. The first stream, the compressor bleed stream, is pressurized and hot (from the inherent heat of compression found within the engine). This high-pressure, high-temperature gas enters a centrifugal compressor after it is first cooled by a heat exchanger using ram air. The bleed air has both its pressure and temperature increased in this initial centrifugal compressor. The high-temperature gas is then directed through a heat exchanger where it is cooled by the second (ram) air stream. Ram air is air from the free stream surrounding the airplane, and at 40,000 ft. altitude can be as low as −40 F or lower. The slightly cooled gas passes through a turbine which shares a common shaft and is driving the previously described compressor. The air is cooled by the energy extraction of the turbine and can now be directed to the cabin. The turbine extracts enough energy from the air to power both the compressor and the Ram air fan. The outlet temperature of the pack can be adjusted by changes to the Ram air flow rate. Higher flow rates will remove more heat from the conditioned air, sending cooler air to the mixing manifold and reducing the overall cabin temperature. Lower Ram air flow rates will send less heat overboard on the Ram air exhaust, which will raise the pack discharge air temperature that enters the mixing manifold.

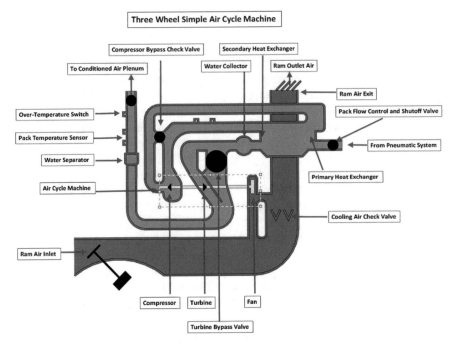

Fig. 4.4 Three-wheel simple ACM [11]

Most of the mix manifold air enters the cabin through larger vents, adjusting the small valve above the individual passenger seat simply increases the local airflow.

The bleed system supplying the air-conditioning packs can use two different compressor tap-off points. A high-pressure line is typically used during taxi and descent when the engine is not at high power. A low-pressure line can provide sufficient pressure and flow during all other engine operating points. These bleed ports take air that has been compressed but ducts it out of the engine prior to combustion and turbine work; therefore it does not contribute to overall thrust, essentially reducing engine efficiency. The 787 has incorporated an electric compressor to more closely match pressurized air requirements to the EC system need. Boeing has claimed a 1–2% improvement in fuel burn at cruise by discontinuing the use of engine bleed and relying on an electrically powered compressors [12]. Many of the health concerns from potentially contaminated air leaving the compressor section of the engine and entering the passenger cabin will be eliminated by adopting these electrically driven compressors to supply pressurized air to the ECS.

The second gas flow stream for the ECS pack is the Ram air circuit. Ram air is cool, external air that can be used as a heat sink due to its low temperature. Ram air derives its name from its ability to extract air from the high-velocity region surrounding the aircraft. The Ram air duct is designed to minimize losses from this airstream and pass it across heat exchanger surfaces with low pressure drop. The third wheel on the three-wheel ACM is the fan driving air in the Ram air circuit

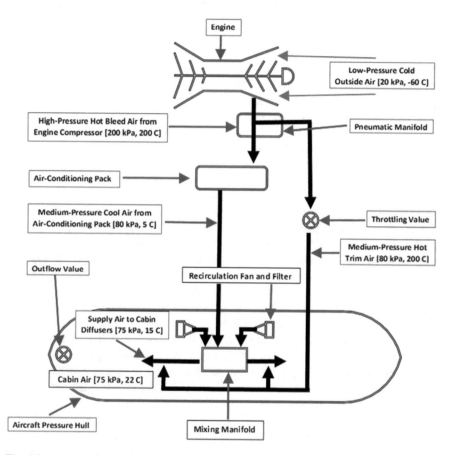

Fig. 4.5 Typical commercial airliner environmental control system

across the heat exchanger and thereby removing heat from the compressed air flow. The Ram air inlet and outlet ducts are carefully sculpted to minimize aircraft drag while allowing free flow of the very-low-pressure external air into the Ram duct.

Figure 4.5 shows the overall environmental control system (ECS) for a typical aircraft and contains nominal values that could be seen during the cruise phase of flight. This shows how the conditioned air leaving the pack is combined with recirculated air from the main cabin. Despite recirculating a significant amount of cabin air, FAA regulations require at least 0.55 lbs./min of fresh air makeup per passenger during the flight [13]. In addition, the CO_2 level cannot be allowed to rise above 0.5% within the cabin. Interesting that this is the same CO_2 level NASA strives to keep as the maximum on the International Space Station.

Most EC systems on commercial aircraft send the very dry, conditioned air, from the pack discharge flow to both the mixing manifold for the main cabin and the cockpit. Therefore, the cockpit is supplied with 100% fresh air makeup from the pack, while the cabin retains some humidity from the respired, recirculated air. While flying at 40,000 ft. the humidity content in the incoming air is extremely low, therefore humidifiers have been added to some cockpit air supply lines on long-haul flights to provide improved aircrew comfort. The ability of air to retain moisture is a function of its temperature, decreasing with lower temperatures. At the -40 F temperatures found in the 30,000–40,000 ft. altitude range, moisture levels are extremely low. This explains the reason passengers can dehydrate so easily, their respiration in the dry cabin environment sends a lot of their individual moisture into the cabin. The primary source of humidity found in commercial aircraft cabins is the respiration from the passengers.

An additional concern for passenger comfort and safety when flying at high altitudes (e.g., 40,000 ft. or higher) comes from the high concentration of ozone at these flight altitudes. Ozone (O_3) is an oxidizing agent that can attack/irritate lung tissue and is found in higher concentrations at high cruising altitudes, especially at high latitudes. The FAR limits O_3 levels to 0.25 ppm instantaneous, and 0.1 ppm over a 3-h period. These levels can be exceeded in flight, leading many EC systems to include ozone converters in their systems. These catalytic systems (similar in geometry to automotive catalytic converters) convert the ozone into diatomic oxygen, making it safe for human breathing [13]. The ozone converters are located in the high-pressure bleed air lines before the air gets to the AC packs.

While many passengers may think the AC system is unavailable on the ground, most airliners rely on an auxiliary power unit (APU) to provide compressed air for these same air-conditioning packs. The APU is meant to operate when the main engines are not running, and local conditioned air is not available for maintaining cabin comfort. The APU typically located near the tail can provide aircraft electrical power, compressed air to operate EC system and compressed air to start the main engines. The APU is a smaller gas turbine that does not provide propulsion, but instead provides compressed air or electricity when the aircraft generators attached to the main engines are not energized. The APU has evolved into the gas turbine device because it can supply high-pressure air and is a high-power-density device (it can fit into a small volume). Because they are smaller gas turbines, they tend to be less efficient than the main engines. However, because they are so much smaller and don't provide propulsion, they also have much lower fuel burn rates compared to the main engines. Hence there is a strong cost incentive to operate the APU vs. the main engines to supply power and air-conditioning, when the plane is stationary, and no propulsive power is required.

While the EC system provides conditioned air to maintain a comfortable cabin temperature and pressure, there is an emergency system for providing oxygen in case the pressurization system or pressure boundary fails. If that happens, the cabin does suddenly lose pressure and equilibrate with the outside air pressure, which could be 40,000 ft. altitude. There is not enough oxygen available in air at this pressure (See Fig. 4.2) for passenger consciousness, so emergency oxygen masks will deploy from compartments above each passenger seat. These oxygen masks are usually supplied by chemically stored oxygen, chlorate candles. Chlorate candles are an efficient long-term storage mechanism for O_2. The chlorate candles will react when ignited to produce gaseous oxygen for the passengers [14]. The pilot will also maneuver the airplane to a lower altitude in the event of a depressurization accident. The chlorate candles have a finite oxygen supply life, on average they will last 10–20 min. Once ignited, the candles cannot be easily extinguished, but must be replaced after one use, or after they have reached their end-of-storage life.

References

1. Fluid Mechanics, White, F., 4th Ed. 1999, McGraw-Hill, ISBN 0–07–069716-7
2. US Naval Flight Surgeon Manual, 3rd Edition.
3. Aviation Medicine in its Preventive Aspects, Fulton, J., Oxford University Press, 1948
4. FAR 135.89, Pilot Requirement use of oxygen.
5. Dressing for Altitude, US Aviation Pressure Suits Wiley Post to Space Shuttle, Dennis Jenkins, 2012, NASA
6. http://www.boeing.com/history/products/model-307-stratoliner.page, retrieved Sep 3, 2017.
7. The Airliner Cabin Environment and the Health of Passengers and Crew, National Research Council, 2002, ISBN 0–309-56770-X
8. https://www.emirates.com/english/flying/our_fleet/emirates_a380/emirates_a380_specifications.aspx retrieved Sep 10, 2017
9. http://www.boeing.com/history/products/777.page retrieved Sep. 10, 2017
10. Aircraft Environmental Control Systems, Dechow, M., Nurcombe, C., HdB Env Chem Vol 4, Part H (2005) Pp 3–24
11. Inside the 747–8 New Environmental Control System, Brasseru, A. Leppert, W., Pradille, A. Boeing Aero, Q1, 2012.
12. 787 No-Bleed Systems: Saving Fuel and Enhancing Operational Efficiencies, Sinnett, M., Boeing Aero, Q4, 2007
13. FAR 25.831 Ventilation,
14. Aviation Maintenance Technician Handbook – Airframe, Vol 1, FAA-H-8083-31, Chap. 16 Cabin Environmental Control Systems.

Chapter 5
Wheels, Brakes, and Landing Gear

Landing gear is frequently one of the last subsystems to be finalized during the aircraft design cycle. The landing gear, their attached wheels which frequently have brakes installed within, then must fit into the space available within either the fuselage or sometimes wing structure. This support structure must be designed to carry the entire aircraft weight, including landing and takeoff dynamic loads. The five fundamental requirements of landing gear include:

1. Keep the aircraft stable while stationary or moving on the ground. This item includes stability during crosswinds, and while turning.
2. Maintain a clearance between the aircraft and the ground during movement. This item occasionally gets abused, as aircraft over-rotate during takeoff and strike the ground near the tail.
3. Limit friction during ground movement. An aircraft weighing over 500,000 lbs. needs to be able to be moved by a ground tug, as well as taxi under its main engine power at low settings.
4. Provide for low drag during takeoff acceleration and maintain safe aircraft clearance during takeoff rotation. The landing system needs to keep low friction during the entire takeoff roll and be stable at takeoff speeds.
5. Absorb the impact and dead weight of the aircraft during landing. The aircraft descent rate and landing weight dictates what shock load will need to be absorbed.

The Wright Brothers solved their need for landing gear by using a simple skid design. They took off from a prepared wooden boardwalk surface at Kitty Hawk and landed on the sand, providing a simple arresting and shock-absorbing mechanism for stopping their "first in flight" aircraft [1].

Within a few short years, new airplane designs incorporated wheels and landing gear into their design. The European designed and built "no. 14 bis," first flew in 1906 and had adopted wheeled landing gear. Glenn Curtiss, the primary competitor to the Wright Brothers in the USA, included wheeled landing gear in his designs by 1908. During World War I (1914–1918) landing gear design seemed to focus on tail dragger configurations [2]. The simplicity of its design meant that the tail dragger

© Springer Nature Switzerland AG 2020
T. Filburn, *Commercial Aviation in the Jet Era and the Systems that Make it Possible*, https://doi.org/10.1007/978-3-030-20111-1_5

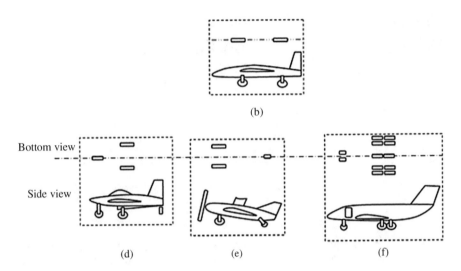

Fig. 5.1 Landing gear type designs

would be one of the early landing gear concepts included in these early aircraft. Figure 5.1 shows the main landing gear types including tail draggers (Fig. 5.1c) as well as the tricycle gear (Fig. 5.1f) favored by most commercial aircraft of the present generation.

Many of the World War I vintage craft had combined the tail dragging design with a clever bungee cord arrangement around the front wheel axle. Incorporating this bungee apparatus allowed the axles that supported these main wheels to deflect 3–4 in. upon landing. This deflection combined with the energy-absorbing capability of the bungee cords wrapped around the axles plus the energy absorbed by pneumatic tires allowed the aircraft to safely land. Of course, the low platform weight (~1500 lbs. empty for a typical WWI fighter aircraft) combined with its average sink rate meant that the impact energy was several orders of magnitude lower than today's transport aircraft (Fig. 5.2).

The tail dragger design shown in Fig. 5.1 continued through interwar years and into WWII. The simplicity of this design helped make it popular during this first half-century of aviation. For this design, the aircraft center of gravity (the imaginary point where the weight of the aircraft is located) will be located between the main landing gear (forward) and the tail wheel or occasionally skid. This insures that the aircraft will meet the five landing gear requirements previously enumerated. While the tail dragger design continued through WWII, aircraft size, weight, and landing speeds continued to increase. In addition, the recognition of the large drag induced by the fixed landing gear remaining in the air stream led to the use of aerodynamic surfaces surrounding the wheel and landing gear strut during the 1920s. Most aircraft employed retractable landing gear during the Second World War to further reduce overall airplane drag. This retraction requirement added complexity to the landing gear geometry and coupled with the increase in aircraft weight meant that new mechanisms would need to be employed to absorb the shock loading of landing.

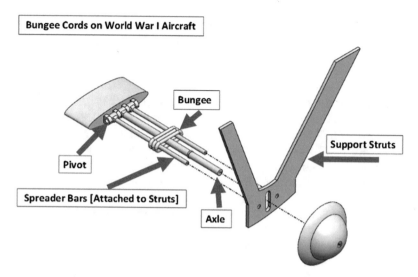

Fig. 5.2 Bungee energy absorber for WWI aircraft

Simply relying on rubber elasticity (bungee cords and pneumatic tires) would not provide sufficient energy absorption anymore.

The Ford Trimotor, an interwar commercial airliner, retained the Tail Dragging design, but used a series of rubber discs located on the main landing gear strut to absorb most of the landing energy. This also used larger tire diameters and tread width to dissipate the impact energy. Unfortunately, pneumatic tires are relatively inefficient devices for this impact load attenuation. In addition, they can provide a rebound energy that may make the aircraft airborne again.

While the tail dragging design stayed for smaller fighter aircraft throughout WWII (with some notable exceptions, e.g., ME262), larger aircraft especially the US bomber force adopted the tricycle design for takeoff and landing. The tricycle design offered both operation and vision improvements but added complexity to the aircraft including the landing gear retraction mechanism and space for a nose wheel. Tail dragging aircraft inherently have a difficult time seeing the runway and potential objects in their path, they are oriented with their nose pointed upward. Tricycle gear keep the wings level and allow the pilots better vision across the runways and taxiways. Tail draggers have a longer takeoff roll as their takeoff roll is initiated with the wings in a high angle of attack, producing higher drag. The tricycle gear provides added stability in crosswind landings and ground maneuvering as the main gear are located behind the center of gravity (Cg) and the aircraft tend to align with the runway due to the large friction load on the main wheels. The aircraft Cg is the imaginary point where the entire load is pressing down to the ground. The Cg can shift fore and aft, plus side-to-side, as loads move within the aircraft. For instance, passengers, luggage, and especially fuel movement and consumption can all move the Cg during operation. While the Cg generally remains close to the aircraft centerline from a port/starboard view, it can adjust significantly in the fore and aft direction. Passengers in some smaller commuter aircraft may have experienced flight attendants asking them to move their seat either fore or aft. This movement is to ensure that the aircraft Cg is properly located for the takeoff limits listed by the aircraft manufacturer.

In addition to the added complexity of the now popular tricycle landing gear, operators need to be especially concerned about Cg and passenger placement. For instance, the ATR72 requires a tail support be installed before passenger loading and unloading [3]. This device prevents the tail from striking the runway if the Cg shifts during the passenger and luggage loading/unloading process. It has pushed operators of this aircraft to load front seat passengers first and unload from the rear of the aircraft. Apparently, some aircraft operators learned this lesson from unnerving passengers by having the tail strike the tarmac while going through the loading process.

The tricycle gear is favored by most commercial airliners, and is the design exclusively used by the present generation of wide-body and single-aisle airliners produced by both Airbus and Boeing. In this configuration, the Cg is located forward of the main gear, but the Cg is closer to the main gear vs. the nose wheel. This puts 80–95% of the aircraft weight to be supported by the main gear and the remaining 5 (minimum load to maintain proper steering) to 20% of the aircraft weight to be held up by the nose wheel or wheels.

The tricycle gear may employ one or two nose wheels and frequently do not employ brakes in the nose wheels. The nose wheel(s) are smaller vs. the main gear because of the lighter load they hold, and their supporting structure does not incorporate the same impact absorbing capability as the main gear. The aircraft descent rate has dropped dramatically by the time the nose wheel contacts the runway. A second nose wheel may be added for higher aircraft loading, or to provide safety and stability in the event of a tire blowout during takeoff or landing.

While the tricycle gear have become the dominant design in civil transport, other designs have found utility in military aircraft. The U-2 used two central main bicycle wheels for takeoff and landing. It used two drop-away pogo legs, with small wheels contacting the runway, located near the wing tips to support the wings during takeoff. The airplane requires a ground chase vehicle to help the pilot bring it in for landing, because it lands at a speed very close to its stall speed, it cannot flare (pitch the nose up) on landing like so many passenger aircraft do. Many of these U-2 design features (fall away pogo supports, thin wings with little stall margin) were included to strip weight and increase the ultimate ceiling of the aircraft.

Many military transports use multiple wheel assemblies to distribute the load over a large footprint. These features allow the planes to be used on unimproved surfaces and still dispatch their loads. The large C-5 cargo plane (the largest plane in the USAF) uses 24 main wheels supporting four struts, while also using a four-wheel nose setup. The more recent C-17 uses fourteen wheels, distributed between two nose wheels, and twelve main wheels that support two main landing gear struts. The long serving C-130 (turboprop) uses two nose wheels, and four main wheels, with two main landing gear struts (Fig. 5.3).

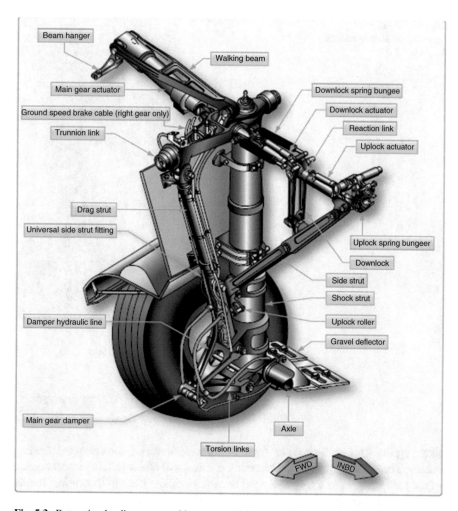

Fig. 5.3 Retracting landing gear used by commercial transport aircraft [4]

Figure 5.4 shows the many parts that are presently used on retracting landing gear found on commercial airliners. The shock strut absorbs the energy of landing without (hopefully) damaging the fuselage or passengers. Many of the other parts are to help in the retraction or deployment phase. In addition, it has features to limit the damage that may occur from FOD on the tarmac.

The key to the energy-absorbing capability of modern landing gear comes from the shock strut (sometimes called Oleo), the main cylinder shown in Fig. 5.4. This device combines an orifice with oil- and gas-filled chambers. Figure 5.5 shows a typical Oleo strut for modern landing gear. The gas is usually compressed above ambient pressure. During landing the descent rate of the plane forces the oil in the lower chamber through a single or multiple orifices. The compression of the strut will also compress the gas phase in the upper chamber of the device. The gas acts

Fig. 5.4 Shock strut
operation to absorb landing
impact

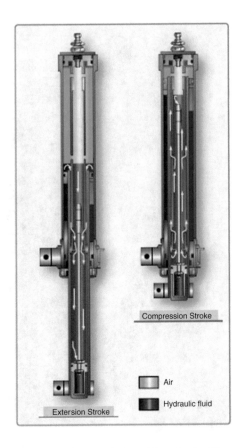

like a spring damper, heating up (heat of compression) and absorbing the landing
energy. The orifice, by limiting fluid transfer rates, will also lessen the recoil bounce
of the compressed gas trying to extend the Oleo cylinder back to its original length
[5]. After landing, the high gas pressure will push the strut down to its fully loaded
but extended state. Limiting this recoil bounce reduces the likelihood of the aircraft
bouncing off the runway and becoming airborne again.

While designed to absorb the energy in normal landings, an over exuberant sink
rate can case the shock strut to bottom out (reach the limit of its vertical stroke)
before the landing energy is dissipated. This can generate a hard landing report, as
well significant loads into the fuselage and the passengers. For large aircraft the
shock strut can compress 10″–24″ or occasionally even more. The larger stroke
values will typically be found on larger, heavier aircraft [2]. Due to the high loads
that need to be supported by the landing gear struts, high-strength materials like
300 M high-strength steel (yield strength ~250 ksi) or titanium alloys are typically
employed.

As aircraft sizes and weights have increased, airplane designers have used several techniques to distribute these loads during landing, as well as taxiing and the
takeoff run. The Boeing 737 uses two main struts with dual wheels for the main

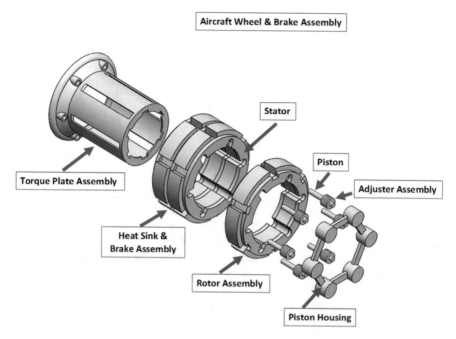

Aircraft Wheel & Brake Assembly

Stator

Piston

Torque Plate Assembly

Adjuster Assembly

Heat Sink & Brake Assembly

Rotor Assembly

Piston Housing

Fig. 5.5 Aircraft wheel and brake assembly

landing gear and a single strut also with dual wheels for the nose gear. This aircraft has grown from 100,000-pound weight when first introduced in 1967 to just under 200,000-pound weight. Boeing has also maintained one weight saving design in its 737 landing gear, it does not have a skin to cover the wheel while housed within the fuselage. Looking up at a 737 after takeoff, one can see two of the main landing gear tires stowed but uncovered in the fuselage.

Larger aircraft like the Boeing 747 and Airbus A380 use multiple struts housed in the fuselage and wing for supporting the aircraft on the ground. The 747 uses four struts each attached to a four-wheel bogie (horizontal beam to connect wheels). The 747 uses a four-wheel arrangement connected to a single strut for its nose gear. The 20 tires on the 747 landing gear can support a vehicle that can weigh up to 970,000 pounds. The A380 uses 22 wheels to support its gross weight (1.25 million pounds). This aircraft uses a dual wheel nose system, and two struts extending from the fuselage plus two struts deployed from the wing section. The fuselage bogie uses six wheels, and the wing bogie relies on four wheels. An important consideration for these ultra-large carriers is the weight distribution onto the runway surface. Not only must the gear support the aircraft during the landing and takeoff operations, but they should be designed such that point loads on the tarmac will not break up the runway surface.

The larger aircraft (747, A380 and others) have a pitch feature incorporated into their landing gear design and operation. This pitch mechanism ensures that the entire bogie will be retracted in that angle, insuring its fit into the wheel bay.

A second aspect to the pitch mechanism is its ability to limit the number of wheels touching the ground at any moment. If all the wheels contacted the ground simultaneously, the large increase in friction could overload the landing gear hardware. Therefore, the pitch mechanism usually puts the rear most wheels in contact with the runway first, and then quickly drops the remaining wheels into contact.

Tires

Aircraft tires have inflation pressures much higher than automobiles (typical 35 psig) and even greater than over the road trucks (100–125 psig). Aircraft tires for regional aircraft can be in the 100–150 psig range, but large, dual-aisle aircraft can have tire inflation pressures over 200 psig. The reason is because the landing gear and tires need to be able to statically hold the entire aircraft weight (Max Takeoff weight, MTOW) where these loads can reach 750,000 pounds for some variants of the 747, and the A380 can reach over 1,000,000 pounds. With a finite number of landing gear struts and wheels, there is a limit to how much this gross weight can be distributed. The large static load coupled with the large shock load of landing have driven design engineers to higher tire pressures.

The conventional rubber pneumatic tire that has worked for the automotive and trucking industry for over a century has been adopted by the airplane industry for several reasons.

1. A pneumatic tire has low rolling resistance, allowing an aircraft to accelerate to takeoff speed in a reasonable takeoff distance, and allows taxi/ramp movement with low power input.
2. A rubber tire has a good compromise in runway/tire friction. This allows large braking forces to be applied through the tire, and reasonable deceleration forces to be transmitted through the aircraft frame.
3. The pneumatic, rubber tire helps the landing gear strut absorb some of the downward landing energy, dissipating it before it is transferred to the fuselage and occupants [6].

Wheels and Brakes

The commercial aircraft wheel assembly has evolved into a common configuration of an aluminum wheel with carbon brakes installed inside the rim. In order to minimize overall landing gear weight, they have adopted aluminum (47,000 psi, yield strength) for the wheel rim. Brake material at the start of the jet era used steel braking material. Some aircraft including Boeing 737 still use this material. However, carbon/carbon matrix composites have become the dominant braking material because of their improved wear characteristics. In addition to their ability to increase

Fig. 5.6 Aircraft brake
assembly after high-speed
RTO, fuse plugs safely
melted and deflated tires

the longevity of brake life, carbon brakes can absorb higher temperatures allowing
them to use smaller brakes to absorb the same braking energy. Figure 5.6 shows a
schematic of a typical wheel and brake assembly for a commercial aircraft. Carbon
brakes began their service on military aircraft in 1990s but transitioned to the commercial arena soon after because of their weight saving, and longer life. They now
dominate the commercial aircraft brake installations.

Aircraft brakes differ from other vehicle designs as they combine multiple rotors
and stators, which in the most advanced design both are made from a carbon/carbon
composite. Instead of a single rotor with friction discs slowing it down (automotive
disc brake) or pads rubbing the inside diameter of the wheel (automotive or alternate
vehicle drum brake), the entire stack is compressed by either hydraulic pressure or
electric brake actuators. In fact, the industry refers to the rotors and stators as heat
sinks as they are designed to absorb the landing energy of even the largest aircraft.
This multiple rotor and stator design generates higher friction loads by having more
surface area in contact during the braking event. This design also allows more even
heat distribution among the brake stack (rotors and stators).

Most ground vehicle brakes are designed to efficiently convect (dump heat to the
surrounding air) heat from the brake surfaces to keep the brake assembly cool and
operating effectively. Aircraft brake assemblies are surrounded by (typically) an
aluminum wheel and a rubber tire. Those materials in the wheel which lose their
strength at relatively low temperatures must be shielded from the brake energy that
is absorbed by the carbon/carbon aircraft brakes. A normal aircraft landing and
braking event can easily see heat sink temperatures reach over 1000 F. Higher loading events (no thrust reversers) can easily double this brake temperature (2000 F
brake temperature). These high brake temperatures, located a fraction of an inch
from the aluminum wheel, have led design engineers to implement wheel thermal
fuse plugs. The fuse plugs are safety devices that prevent the tire pressure from
reaching dangerous levels due to an advertent thermal load. This fuse plug is a
eutectic material, that will melt at temperatures below the point at which the tire or

rim will weaken [7] due to elevated temperatures. This fuse plug saves the high-pressure tire from explosively detaching from the rim and provides added safety for ramp and airline personnel who may have to work in the landing gear area.

The tricycle landing gear type used by most transport planes allows a higher braking force to be applied, without the front end nosing over when forcefully applied [8]. The tail dragger configuration needs to limit main gear braking force to prevent the nose from hitting the runway during braking events. This is another example of why the tricycle gear type has dominated the commercial aviation market.

The most energetic braking event comes from the potential of a rejected takeoff. A rejected takeoff (RTO) occurs when the pilot consciously stops the takeoff roll. During the takeoff acceleration the pilot will take note of the aircraft passing through several speed points. The first, V_1, is the when the plane has achieved sufficient velocity that the takeoff should not be aborted despite any anomaly (e.g., flat tire, engine failure etc.). V_1 is the point where the aircraft has too much velocity and too little runway left to abort the takeoff safely. V_2 is the velocity at which the aircraft can safely climb with only one engine operating. The third speed, V_R, is the speed at which the nose can be pitched up, and the aircraft can start to leave the ground.

The most difficult challenge for the brake assembly is the requirement to handle a rejected takeoff at V_1, this is the max velocity at which the aircraft may safely abort the takeoff. The RTO requirement stems from the need to stop a max loaded airplane at its highest velocity, to further challenge the braking system, the brakes will be tested at their minimum thickness (max wear), and for the FAA demonstration test, the thrust reversers cannot be deployed. The max wear on the brakes just demonstrates that all brakes (new or old) will have sufficient stopping power to stop the plane. Keeping the thrust reversers off recognizes that one significant contributor to a rejected takeoff can be loss of an engine on a twin-engine plane. This same one-engine-out scenario would produce extreme aircraft yawing, if the thrust reversers were deployed. While many in the traveling public have not experienced an RTO, Boeing estimates that 1 out of every 3000 takeoff attempts results in an RTO. This means that a pilot flying international routes may see one of these events every 20 years. However, pilots flying much shorter feeder routes may experience these events every 7 years [9].

The total energy that needs to be absorbed by the brakes in an RTO scenario can range from 40,000 Hp for a single-aisle aircraft to 250,000 Hp for a fully loaded wide-body aircraft [6]. A typical automotive SUV has a ~300 Hp engine. So, this braking event can exceed the energy produced by 800 cars in tandem. Brake temperatures can reach over 3500 F during this extreme test.

Figure 5.6 shows a wheel and brake assembly after a high-energy braking event, at a velocity close to V_1. This picture demonstrates the enormous amount of energy that is held within the brake pads after this braking event (note the brake pads usually can be seen as bright red objects inside the wheel assembly). In Fig. 5.6, the DC-10 shown actually overran the runway at LAX due to other tire problems. The FAA recognizes that damage to the wheel and tires will probably occur as a result of a high-speed RTO. The agency requires that no fire appear above the top of the tire for the first 5 min after one of these braking events. This 5-min time period will give emergency response personnel time to deploy and prevent fire from spreading to other (occupied sections) of the aircraft [10].

Nose wheel steering is generally available on most commercial aircraft. This allows the pilot to effectively steer the aircraft around the ramps, taxiways, and parking at the gate. Another method available for directing the aircraft is through differential braking. The pilot can apply braking force to the main aircraft wheels in an uneven fashion. This method allows the aircraft to reach even tighter turning radii, effectively pinning the aircraft on the brake/wheel/landing gear strut through which the brake force is applied. This type of maneuver also places a significant load on the tire, brake, and landing gear strut that is locked via the brake application.

References

1. https://www.nps.gov/articles/wrightflyer.html, retrieved 9/3/17
2. Aircraft Landing Gear Design: Principles and Practices, Currey, N., AIAA 1988, ISBN 0930403-4 I-X
3. Questions raised over Airworthiness of ATR-72 Aircraft, Dawn, May, 31, 2015, Zulqernain Tahir
4. Aviation Technician Maintenance Handbook – Airframe, Vol 2, FAA-H-8083-31, 2012
5. Sport Aviation, April 2105, Oleos, Busch, M.
6. Landing Gear Design, Chapman & Hall LTd, Conway, H. 1958
7. J. Aircraft, Vol. 5, No 1, Jan-Feb 1968, Stanton, G. "New Designs for Commercial Aircraft Wheels and Brakes"
8. AMT Airframe Handbook Volume 2, (FAA-8083-31), Chapter 13 Aircraft Landing Gear Systems.
9. Rejected Takeoff Studies, Boeing Aero Magazine, No. 11, Q3, 2000
10. FAA Advisory Circular AC 25.735-1 Brakes and Braking Systems Certification Tests and Analysis. 4/10/02

Chapter 6
Fuel Systems

The aircraft fuel system has evolved from its basic and humble beginnings on the Wright flyer. The Wright flyer relied on a simple configuration of the fuel being gravity fed from the gas tank situated above the engine. The fuel system needs to be an early and important design consideration because of its immediate impact on the propulsion system operation coupled with its essentially linear impact on flight endurance.

The Wright brothers located their small 1.5 gasoline tank on a strut several feet above the engine location. Their fuel system then used an in-line regulating valve to adjust fuel flow before takeoff. This fuel flow remained unchangeable during flight (except for the very slight decrease in flow due to the small liquid level change in the tank as fuel was consumed). Their design allowed the fuel to be vaporized by contacting the outer surface of the crankcase and be mixed with the incoming combustion air. Because of their choice for an internal combustion engine, they needed a fuel vapor-air mix entering the cylinder before ignition.

The Wright flyer incorporated several design features that would be hallmarks of fuel systems all the way through the jet age. The overall fuel tank capacity will set flight endurance, except for military aircraft which can employ in-flight refueling. The fuel tank location is important for flight control as it changes the aircrafts center of gravity (Cg), and this Cg can move as fuel is consumed during flight. The Wright flyer did not incorporate several features that present-day jet engine fuel systems operate with. The first fuel tank inerting was not a consideration during the Wright days but is now regularly employed on military aircraft and has become more widespread on commercial craft. The second system now widespread but not seen on the Wright flyer is using the fuel as a heat sink (energy dump) for various airplane systems.

Early aviators started with gliders, but adopted internal combustion, spark ignited engines to provide endurance and eliminate flight being dictated by the vagaries of the wind. These engines rotated the propellers, which produced thrust and generated lift due to the airflow over the wings. The Wright Brothers and others in these very first airplanes simply relied on the gasoline mixtures available at the time, as the

© Springer Nature Switzerland AG 2020
T. Filburn, *Commercial Aviation in the Jet Era and the Systems that Make it Possible*, https://doi.org/10.1007/978-3-030-20111-1_6

limited number of automobiles meant that no large-scale refueling system was in place yet. As airplanes ventured higher, for longer flight times, and as everything aviation related depended on weight, aviation fuels began to differ from the blends used for land-based (automotive) applications.

Details about the type of fuel used by the Wright brothers cannot be reliably determined, but most accounts list them using gasoline from a local boatyard. The type of fuel used is especially important for aviation, which unlike ground transport, can result in more injurious results if the engine doesn't turn. A stalled auto leaves a vehicle stranded by the side of the road, a nonfunctioning airplane engine can lead to catastrophic results if the fuel does not operate as planned. With such a premium on weight, fuel energy density is important, along with its ability to combust inside the engine, and it must be able to stay a liquid without large viscosity swings despite large temperature variations (~ −40 up to 125 F).

The early aviation engines operated on gasoline, which differs from kerosene, and later jet engine fuels. Gasoline tends to have higher volatility vs. the lower vapor pressure present-day fuels like kerosene and jet A. Engine "Knock," the premature detonation of the fuel, was a problem in these early aviation and automotive engines which generated research to understand the cause and potential prevention during World War I. Predetonation or knocking can severely affect the performance and power output of any internal combustion engine.

The First World War delivered many advances in aviation technology, including engine power, powerplant thrust, aircraft endurance, and ceiling. While investigators could not and did not alter the fundamental flammability of gasoline nor its energy density, it remained the preferred fuel during this time. Not surprisingly, the frequent exchange of gunfire plus the numerous flammable components on these early aircraft led to many instances of fires on-board the aircraft during the hostilities. US aviators were at a disadvantage, as their leadership refused to endorse the use of parachutes, believing they would reduce the pilot's will to fight. Parachutes were allowed and put to use by the other Allied Air Forces [1].

Aviation gas is similar to automotive gasoline, it is one of the lighter and more volatile fractions of oil as shown in Fig. 6.1. Aviation gas (Avgas) has additional requirements beyond those needed for the automotive industry. Avgas needs to meet stringent reliability and usability standards including, vapor pressure, non-freezing, plus it needs to prevent predetonation (knocking). To prevent predetonation tetraethyl lead (TEL) began to be used in gasoline as early as 1927, when the US Navy used it in their new 410 HP Pratt & Whitney Wasp engine [2]. The automotive industry began using lead as a gasoline additive in 1922 for the same anti-knock (higher octane rating) properties, it was discontinued as an additive in automotive gasoline beginning in 1974. It was not the danger from lead pollution that resulted in it being discontinued, but rather to protect the catalytic converters that were being introduced to reduce automotive CO and hydrocarbon emissions. The first half of the 1970s saw the US institute dramatic air pollution control rules, including automotive emissions. Despite a total ban on the use of lead in automotive gasoline that dates back to 1995, Avgas still predominantly relies on TEL to increase its octane

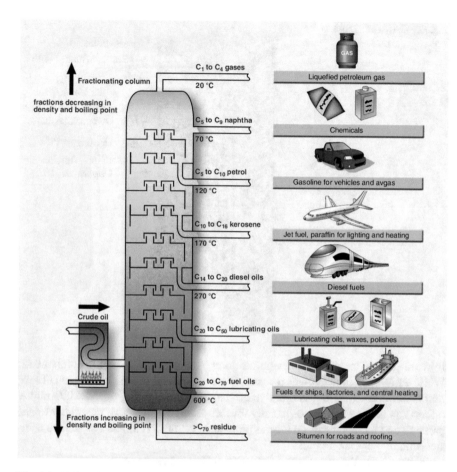

Fig. 6.1 Oil fraction into various constituents

rating and prevent predetonation. While automotive engine designers figured out how to work with fuels without TEL, the aviation industry has not [3].

The airplane industry used the interwar period (1919–1938) to develop improved fuel system components and increase operational reliability. The driver for these performance improvements included the higher power, longer endurance platforms that were being designed and built. In addition, the susceptibility to fire, the flammability of AvGas and the need to protect combat aircraft played a role in these design improvements. During this time, government agencies also began to develop standards on the makeup and performance of AvGas.

One improvement made to the fuel system came from the decision to use 100 Octane fuel for the US Army Air Corps in 1938. The oil and automotive industry developed the Octane rating system as a measurement of gasoline's tendency to predetonate. A 100 Octane rating, the highest rating, meant the fuel had the greatest resistance to knocking [2]. Predetonation can dramatically drop the power output

Fig. 6.2 Self-sealing fuel tank

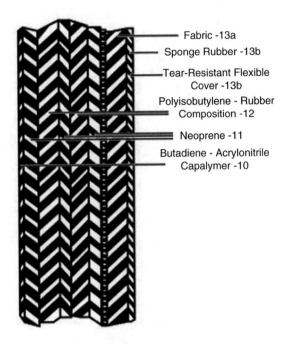

Fabric -13a
Sponge Rubber -13b
Tear-Resistant Flexible Cover -13b
Polyisobutylene - Rubber Composition -12
Neoprene -11
Butadiene - Acrylonitrile Capalymer -10

from an engine, as well contributing to long-term damage to engine components. While anti-knock is important for cars, it can be critical in aviation, especially for military aircraft. Regular automotive gas has an Octane rating of 89, while higher ratings can also be bought, especially for turbocharged car where the higher intake pressure and temperature can lead to premature detonation unless the vehicle operates with higher Octane fuel.

In addition to the adoption of AvGas (100 Octane) as the predominant fuel type, the industry applied rigorous standards to its properties, production, and improved AvGas storage technology during the interwar years. The major US rubber companies investigated methods to produce self-sealing fuel tanks during this same interwar time period. Their composite solution, shown in Fig. 6.2, relies on a soft-rubber sealant that swells when exposed to fuel. This rubber layer is sandwiched between two layers of material that will not allow fuel to leak. If a bullet penetrates the composite lay-up, the middle layer now contacting fuel will swell and stop the leakage of fuel.

As the range and size of aircraft increased, their fuel tanks naturally grew larger and more sophisticated. Jet engines use a less volatile fuel type compared to the previous generation piston engines (close to kerosene). However, the low efficiency of the early commercial jet engines vs. the final piston engine-propeller combination available in the late 1950s also dictated the need for even larger fuel tanks for these early jet-powered aircraft.

Both ground vehicles and airplanes measure their fuel tank size in volumes. However, the high premium placed on weight in the aerospace industry pushes airlines to describe their fuel loading by mass. Additionally, commercial aircraft need

Table 6.1 Commercial airliner fuel tank capacity (typical)

OEM	Model	Fuel tank volume (L)	Max fuel mass (tonnes)	Range (km)
Embraer [4]	E175	11,625	9	4074
Boeing [5]	737–500	20,100	16	4400
Airbus [6]	A380	323,546	254	15,200

Table 6.2 Jet A vs. AvGas 100 properties [8, 9]

Property	Jet A	AvGas 100
Fuel boiling range (C)	150–265	45–145
Freeze point (C)	−40	−60
Density g/mL (at 16 C)	0.81	0.72
Net heating value (kj/kg)	43,140	42,800

to keep track of their Cg, which means that fuel load on a mass basis makes for quicker calculations of Cg. Table 6.1 shows the fuel tank (volume and mass) capacity for various scales of commercial jet aircraft including commuter size, single aisle, and wide-body.

While high octane Avgas became the predominant fuel of the piston-powered aircraft, jet fuel has evolved through the decades. The commercial airplane builders plus operators and their engine makers have settled on Jet A in the USA. A close variant Jet A-1 can be found in the USA but has wider acceptance worldwide. These jet fuels are similar to kerosene in their distillation point, volatility, and combustion properties. Jet B has also been developed which is a blend of kerosene and gasoline. The Jet B mixture produces a lower freezing point (−58 °F vs. −40 °F for Jet A), making it a popular choice in Alaska and Canada [7]. Freeze point can also be an important criterion on long flights (like Polar flights) where the fuel temperature can approach the outside air temperature (−55 C, −67 F) due to the multiple hour flight time. If the fuel gets too cold, it can congeal on screens designed to protect the fuel system, producing a total blockage of fuel flow and thus generate an engine-out condition. Even a single engine flame out can be a very scary scenario when both engines are running on the same fuel source and most commercial aircraft now rely on only two engines. Table 6.2 shows a comparison of the properties of AvGas 100 and Jet A Fuels.

Table 6.2 demonstrates that Jet A (and its international variant Jet A-1) tend to be less volatile than AvGas but with nearly identical energy (net heating) values. A key parameter for any aviation fuel is its density (which will change with temperature). While fuel tanks are measured in volume, fuel energy is based on mass. Therefore, aircraft range can be strongly impacted by the ambient temperature when fueling the aircraft. High air temperatures reduce fuel density, thereby reducing aircraft range. In a double constraint, these same high temperatures will also impact engine operating performance, reducing thrust and in extreme cases generate commercial flight cancellations [10]. The thrust from a gas turbine engine depends on the mass of air entering (and exiting), and aircraft lift is dependent on air density. Air density goes down as air temperature goes up. Air temperatures above 118 F in Phoenix during the summer of 2017 generated reduced flight loads or even flight eliminations. These flight cancellations arose because certain airplane types had not been certified to takeoff under those high air temperature (low air density) conditions. These conditions led to flight cancellations because their aircraft had not been certified by the FAA to takeoff under those extreme temperature conditions.

Commercial Jet Aircraft Fuel System Requirements and Design

As listed in Table 6.1 a significant volume of fuel needs to be stored in commercial aircraft, not surprisingly a larger amount for the wide-body intercontinental flights. The Original Equipment Manufacturers (OEM), Airbus, Boeing, Embraer, and others have elected to use the volume inside the wings for a significant fraction of their mission fuel (some designs may use wing tanks exclusively). Larger aircraft with longer mission profiles (and therefore greater fuel requirements) tend to rely on center fuel tanks in the region below the fuselage between the wings as another storage location. The combination of segregated fuel tanks (wings and central body) along with the consumption of fuel from a fixed location fuel during flight will produce a shift in the aircraft Cg during a flight. Transfer pumps in the fuel system can adjust the remaining fuel storage to minimize or eliminate this Cg shift and keep the aircraft trimmed for minimal drag. In general, the center of lift does not change during cruise, but Cg shifts during this same time can generate a nose-up or nose-down trend.

During the aircraft design phase, the fuel tank location must be selected in concert with the engine location. Nose or tail tank locations will create long fuel lines for wing-mounted engines. In addition, these same nose or tail locations would generate larger Cg shifts because of fuel consumption during flight Fig. 6.3.

Another requirement for fuel tank location requires that they keep out of the engine rotor burst zones. Rotor bursts are regions that have the potential for high-speed components (compressor and turbine sections) within the engine to impact. This requirement stems from the potential for a significant engine failure to spray high-speed and high-energy debris. Engines cannot be designed to retain these components within the engine housing. Figure 6.4 shows an engine with its radial rotor burst zones, and the fuel tanks located to avoid these potential regions where high-energy engine parts could potentially strike.

The fuel system provides a storage place, and continuous flow of fuel to the propulsion engines (and to the Auxiliary Power Unit, APU, if the aircraft is so equipped). As previously described, the fuel system must be capable of moving fuel fore and aft to maintain the aircraft Cg position. In addition, port/starboard fuel movement may be required, especially in an engine-out condition, to keep the aircraft laterally stable. Not often used, but the fuel system must be able to jettison fuel in the event that the aircraft needs to land shortly after takeoff (medical emergency, engine failure

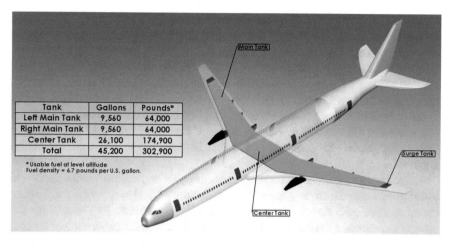

Tank	Gallons	Pounds*
Left Main Tank	9,560	64,000
Right Main Tank	9,560	64,000
Center Tank	26,100	174,900
Total	45,200	302,900

* Usable fuel at level altitude
Fuel density = 6.7 pounds per U.S. gallon.

Fig. 6.3 Typical commercial aircraft fuel tank locations

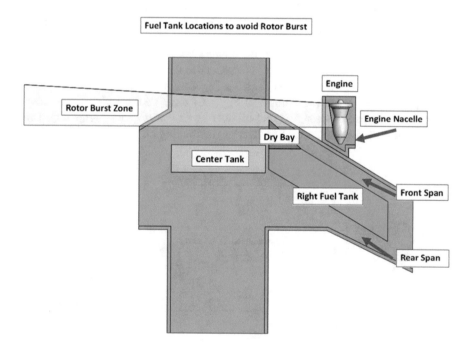

Fig. 6.4 Fuel tank locations to avoid rotor burst zones

on takeoff, etc.). Max takeoff weights almost always exceed max landing weight as the dynamic landing gear load is increased by the combined aircraft deadweight coupled with its sink rate during landing. The fuel jettison system provides a way to send fuel overboard, without having to wait for it to be burned in the engines [11].

The fuel system needs to provide a method to meter the fuel to the engines based on the thrust setting initiated by the pilot. Today's high-bypass turbofan engines now rely on very high compression ratios coupled with the high bypass ratios to achieve their higher fuel efficiency (thrust-specific fuel consumption, TSFC). These engines routinely reach air pressures in the combustion chamber up to 1000 psig, which obviously means that the fuel injection pressure must be 1000 psig or higher in order for fuel to reach the combustors. The fuel system needs to increase the fuel pressure from the very low ambient pressure of the vented fuel tanks (~2.4 psia) up to pressures above the combustion chamber in order to get sufficient flow to the combustor.

Large commercial transport aircraft generally rely on multiple pumps located in series to generate the large pressure rise needed for the fuel to enter the combustion chamber. Figure 6.5 shows a general schematic of the *airframe* fuel system and the other components in the fuel line. The airframe fuel system is that portion of the fuel system that collects the fuel from the various regions of the fuel tanks and delivers it to each engine where the resident fuel systems at each engine location finishes pressurizing and distributing the fuel inside the engine. These airframe fuel systems do not achieve a high pressure and are merely meant to collect, transfer, or send fuel overboard if required.

Figure 6.6 shows a schematic of the Engine Fuel system, housed within each engine location. This engine fuel system generates the very high pressures needed to inject fuel into the combustion chamber (typically through a positive displacement pump). The system (as shown in Fig. 6.6) starts with a low-pressure boost pump and then passes through a fuel-oil heat exchanger (to cool engine oil). This device transfers the engine friction energy (lubricating oil heating) back into the

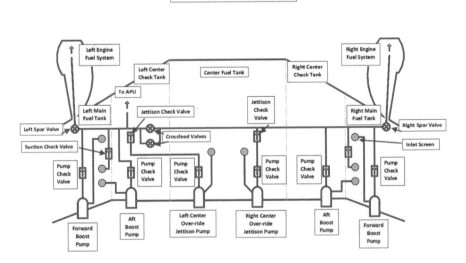

Fig. 6.5 Airframe fuel system schematic [12]

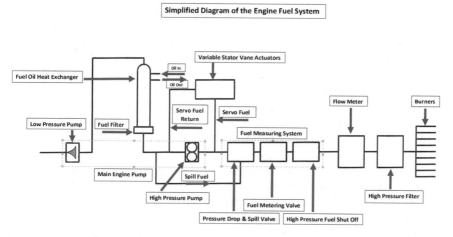

Fig. 6.6 Engine fuel system [12]

engine for combustion and improved fuel economy. Downstream of the heat exchanger is a fuel filter to protect the small diameter tubing and the very small passages of the burner nozzles. The high-pressure (positive displacement) pump provides a significant pressure rise for the fuel, some of which is diverted to the variable stator vane actuator system. This device uses "fueldraulics" (high-pressure fuel as a hydraulic fluid) to position stator vanes within the compressor section of the engine. These stator vanes help to improve the aerodynamic performance of the compressor at the widely varying airflows that occur during takeoff, cruise, and descent thrust conditions. This system adjusts the position of the stator vanes (those stationary devices in front of rotating compressor sections) to optimize the direction of the air entering the downstream compressor section. The remaining fuel that is heading to the engines passes through the fuel metering valve, the high-pressure (HP) fuel shutoff valve, the flow meter, and finally the HP filter before entering the nozzles in the engine combustion chamber. The fuel nozzles are designed to atomize the liquid fuel to achieve complete combustion of these droplets in the low residence time available in the combustor region. The nozzles now have added requirements, such as minimizing pollutant formation including NO_X and CO.

Fuel tank inerting has been a technology that has been used by military aircraft for over four decades. Inerting of fuel tanks relates to the fundamentals of combustion and the necessary level of oxygen required to support combustion. Fuel tank inerting generally relies on reducing the oxygen concentration in the gas space above the liquid fuel. Fuel combustion occurs in the vapor phase and is greatly resisted when the fuel is liquid. The military uses this technology in various subsystem designs on board its larger transport aircraft (C-5, C-17, C-130) as well as its smaller combat aircraft. All four variants listed (three transport types and one combat aircraft) rely on different inerting/fire protection technologies, which also cover the majority of systems used by commercial aircraft and will be described below.

The US Air Force (USAF) C-5 transport aircraft uses tanks of liquid nitrogen stored on-board the aircraft which can easily be heated into gaseous nitrogen and delivered to the fuel tank ullage space (gaseous region above the liquid fuel). The nitrogen dilutes the oxygen concentration above the fuel tanks, dramatically reducing the chances of fire or explosion within the tanks. Keeping nitrogen as a liquid greatly reduces the volume (800:1 density difference between nitrogen's liquid and gaseous state) required to store it on-board the aircraft. However, using liquid nitrogen introduces a logistics issue, as the tanks can only be refilled by airfields that have liquid nitrogen stored or available for delivery. The very large C-5 (largest aircraft in the USAF) generally operates as a strategic airlifter from larger airfields, with the infrastructure in place to support liquid nitrogen replenishment.

The latest USAF cargo plane, the C-17, uses an on-board, inert gas generating system (OBIGGS) to produce gaseous nitrogen for fuel tank inerting. The OBIGGS system uses thousands of hollow fiber membranes (HFM) to separate oxygen from nitrogen with air as the feedstock. The system relies on high-pressure air taken from the compressor section of the engine. The air is cooled and flows inside the HFMs where the membrane walls have a higher affinity for oxygen, allowing it to diffuse through the walls more rapidly. With these characteristics, the gas stream flow exiting the tubes of the HFM will begin to reach higher concentrations of nitrogen, while the gas outside the membrane will be oxygen-rich. The system disposes of the oxygen-rich gas stream overboard and uses the nitrogen-rich stream (nitrogen enriched air, NEA) to provide inert gas for the ullage space above the liquid fuel. The military uses a value of 9% oxygen as their max safety limit to prevent fires/explosions in fuel tanks (vs. the ~21% oxygen concentration found in air at any altitude). More recent research has raised this limit to 10%, which is the inert gas/oxygen depletion level used by the commercial aircraft suppliers (Fig. 6.7).

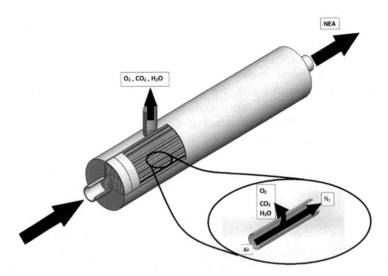

Fig 6.7 HFM system for generating oxygen-depleted air to be used for fuel tank inerting

The USAF relies on open cell foam within the fuel tanks of some C-130 planes to reduce the extent of fuel vapor in the ullage space, thereby greatly reducing the chance for fires or explosions. The foam breaks up the continuous gaseous region above the fuel and limits the propagation of a flame front within the tank. The foam is a parasitic loss, in that it takes up space that could be occupied by fuel. The foam needs to extend throughout the tank region to provide safety during all fuel fill levels (nearly full to nearly empty).

Finally, older model US fighter aircraft (e.g., F-16) relied on Halon to prevent fires in their fuel tanks. Halon is no longer produced due to a worldwide treaty banning its production. This ban arose due to Halons propensity for ozone depletion. As the aircraft carried only a finite mass of Halon, it was designed to be used as the aircraft was entering hostile airspace and could provide only a limited protection period before the Halon was exhausted. Newer model fighter aircraft have switched to OBIGGS so that it can provide fuel tank protection during the entire mission.

The electronic engine control systems used by both military and commercial engines generate a significant amount of internal heat from the large amount of electronics housed within the enclosure that houses this equipment. This heat must be removed in order for the engine control systems to continue to operate properly. Commercial aircraft almost universally use air cooling to take away the heat energy produced by their electronic engine control systems (Full Authority Digital Engine Control, FADEC). However, US fighters also use their fuel system to cool these electronic control boxes. Military engines generally have higher heat loads due to their higher performance requirements, and therefore benefit from a liquid (fuel) coolant. Liquids can absorb more thermal energy vs. gases for two reasons, but both related to their inherent density increase compared to gases. The first reason is because they will have a higher mass flux near the surface to be cooled. This increase allows greater heat transfer to the fluid. The second reason is because the liquid can absorb more heat due to its higher density.

The FADEC provides input for the fuel metering system, which ultimately controls the engine thrust. The fuel metering system provides control for the engine performance. A change in engine thrust (from the cockpit) can only be achieved by a change in fuel flow. Higher fuel flow equals higher thrust (up to the engine limit).

Finally, the fuel system needs to provide fuel to the APU (auxiliary power unit). The APU is (typically) a smaller gas turbine engine located within the fuselage for auxiliary and sometimes emergency power. It is designed to provide electricity and frequently pressurized air for various components when the main engines are inoperative. The APU can usually operate the air-conditioning system, hydraulic system, or other systems when the main engines and ground power are not available.

References

1. http://www.thehistoryreader.com/modern-history/parachutes-world-war-1/, retrieved 12/21/2017
2. A short history of aviation gasoline development, 1903-1980, Alexander Ogston, The Aeronautical Journal, Vol. 85, issue 850, 1981, pp 441–450

3. https://archive.epa.gov/epa/aboutepa/epa-requires-phase-out-lead-all-grades-gasoline.html, retrieved 12/22/2017
4. https://www.embraercommercialaviation.com/commercial-jets/e175/ retrieved 1/1/2018
5. http://www.boeing.com/resources/boeingdotcom/company/about_bca/startup/pdf/historical/737-classic-passenger.pdf retrieved 1/1/2018
6. AIRBUS A380 Aircraft Characteristics, Aircraft and Maintenance Planning, Rev 01/16
7. FAA Aviation Maintenance Technician Handbook, Volume 2, Chapter 14, Aircraft Fuel System. (FAA-H-8083-31)
8. Advancements in Gas Turbine Fuels from 1943 to 2005, T. Edwards, Journal of Engineering for Gas Turbines and Power, January 2007, Vol 129, pp 13–20
9. Aviation Fuel Properties, Coordinating Research Council Report No. 530, 1983
10. Here's Why Planes Can't Take off when it's too hot, A. Jenkins, Fortune, June 20, 2017
11. Aircraft Fuel Systems, R. Langton, C. Clark, M. Hewitt, L. Richards, John Wiley, 2009
12. http://lessonslearned.faa.gov/ll_main.cfm?TabID=3&LLID=79&LLTypeID=2, retrieved 2/4/18

Chapter 7
Instruments and Sensors

Aircraft today rely on a variety of sensors including input from satellites to provide information on their altitude, direction, location, and speed. However, the early pioneers had to develop these instruments to make air travel the safest mode of movement today. The Wright Brothers, the inaugural flight engineers (along with being the first pilots and aircraft designers), relied on the most rudimentary of instruments. Their open air "cockpit" allowed them full visual access to their surroundings as well as auditory monitoring of their propeller and powerplant. In addition, their "cockpit" (they were laying down on the wing) had a stop watch, engine tachometer, and anemometer (for measuring air velocity) [1].

The Wright brothers initiated the early practice of flying exclusively during daylight, when the weather was good. This allowed the pilots in these open cockpit aircraft to continue to rely on external visual cues as part of their instrument suite. The generally broad visibility in the open cockpit provided navigation input, they could and did follow large landmarks (rivers, train tracks, etc.), plus they routinely flew low enough and slow enough that they could read town names on railroad stations. These same open cockpits allowed the pilots to gain reference to bank angle, whether they were climbing or descending, and approximate altitude. Starting with the Wright flyer and continuing to the modern jet airliner, a pilot generally needed three types of information, status of their powerplant, status of the aircraft, and their location plus navigation details.

During the first decade of heavier-than-air powered flight, aircraft instruments had not expanded much beyond the original two instruments included in the Wright flyer, an engine revolution counter and air speed indicator. The engine revolution counter was especially important in these early days as a potential warning of impending engine trouble. The low reliability of these early power plants meant that a method to determine engine health would allow a pilot time to choose a landing site vs. having one imposed on him. Air speed indicators were important because of the small margins available between aircraft stall speed (lose of lift due to low speed) and damage to the flimsy (by today's standards) aircraft (due to overspeed). Finally, the altimeter became important as the aircraft reached higher altitudes.

© Springer Nature Switzerland AG 2020
T. Filburn, *Commercial Aviation in the Jet Era and the Systems that Make it Possible*, https://doi.org/10.1007/978-3-030-20111-1_7

Roland Garros set the then world altitude record of >18,000 ft. in his French monoplane in 1912 [2]. Many of these altimeters and other early instruments had their roots in ballooning and other aerial industries, the large pulsation environment inherent in the engines of the day, made the task of correctly reading the various data points much more difficult than their ballooning roots which tended toward much lower levels of vibration.

At the same time, the nascent automotive industry also had a need for an engine revolution indicator and these instruments were soon adopted for aeronautical use with little modification. They were already designed for a harsh engine vibration environment, but they did need to switch to lower weight alloys for the aircraft industry. These early engine revolution counters or tachometer instruments generally relied on centrifugal indicators with spring-constrained revolving weights. As the engine speed increased a horizontal indicator linked to a pointer moved and provided output.

The Wright flyer used a Richard, pinwheel type, anemometer (see Fig. 7.1) [3] to record the wind speed during the flights. These early low-speed aircraft were dangerously close to their stall speed, so an accurate airspeed measurement would be critical. However, the paddlewheel device used by the Wright flyer did not last long, as it added significant drag and did not provide the accuracy sought by these early aviators.

Fig. 7.1 Pinwheel-type anemometer held by Wright Bros (used with permission, Dan Patterson)

Fig. 7.2 Simple U tube manometer for determining air speed

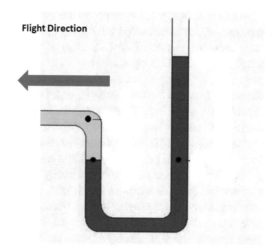

Flight Direction

Fig. 7.3 Pitot probe as proposed by Darwin to measure velocity [4]

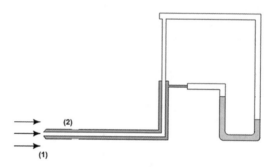

(2)

(1)

By 1913 academics and aircraft designers were focused on accurate measurements of airspeed because of its critical importance in maintaining controlled flight. The community started to move from the vane-type devices like that shown in Fig. 7.1, and instead began to rely on the pressure difference between the ambient pressure and the higher pressure created by the plane's velocity (dynamic head). A simple U tube glass device filled with colored water like that shown in Fig. 7.2 was initially tried, but only produced small differences in liquid level. The small difference in liquid height, the inherent vibration imparted from the engine, the pitch, roll and yaw of the aircraft all made it very difficult to determine velocities accurately.

A more accurate device shown in Fig. 7.3 was first suggested by Horace Darwin, chairman of the Cambridge Instrument Co., also in 1913. This enhanced device provided a more protected static port, and a more streamlined dynamic (pitot) port. A fundamental principle of fluid flow (Bernoulli's equation) can be used to derive the formula for velocity in the Pitot probe of Fig. 7.3.

$$V = [2*(P_0 - P_s)/\rho]^{1/2} \qquad (7.1)$$

In this equation, V is the velocity of the craft (airplane). P_0 is called the stagnation pressure, which is measured through the port aimed in the flow direction. P_s is the static pressure, measured by the port(s) perpendicular to the flow direction. ρ is the density of the air. This equation is valid as long as the probe is pointed in the direction of flight. The height of the fluid column is linearly proportional to the pressure difference between the two ports $(P_0–P_s)$. For a plane velocity of 70 mph, one could expect about a 2 ¼ inch difference in the height of a water column between the two ports. This is large enough that a reasonably sized device could be constructed to read this pressure difference and convert it into a measurement for forward velocity. This ~2-in. column height is also large enough that reasonable gradations can be used to provide some range to the velocity measurement. Not surprisingly, aircraft acceleration or rapid changes in direction could easily upset this liquid column, meaning it would not provide accurate results until the liquid achieved a new static state. It is interesting that this exact same principle (Bernoulli equation) and the measuring of the difference between static and stagnation pressure is still used today to measure aircraft speed. However today, pressure transducers monitor the differential pressure, vs. measuring the height of a liquid column, this provides more reliability, accuracy and does not suffer instability from a liquid column sloshing back and forth.

About this same time, aircraft researchers became interested in inclinometers, to measure the angle of the aircraft with respect to the ground. It was considered the third most important type of instrument (after tachometer and air speed). While early pilots thought they could deduce whether their craft was climbing or diving, their senses were usually grossly inaccurate. The inclinometer would provide a direct and accurate indication of the aircraft angle relative to horizontal. Early attempts to achieve an indication of aircraft angle vs. a horizontal plane simply relied on a fluid contained in a curved tube, with gradations marked for positive or negative angles relative to horizontal. Once again, the aircraft movement (banking, turning as well accelerating and deceleration) all played havoc with these early and simple inclinometer instruments, making their indications unreliable.

The period from July 1914 to November 1918 would provide large increases in airplane performance, more reliable sensors and instruments but the disparate number of producers in numerous different countries hampered any tendency to standardize the types of instruments located in the cockpit. Even so, the very large increase in the number and types of aircraft generated an equal increase in demand for aircraft instruments. It is estimated that over 100,000 aircraft of all types (fighters, bombers and reconnaissance) were produced just by the French and British (the two largest producers) during the war [5]. Figure 7.4 shows a typical cockpit instrument arrangement as the end of the "Great War" approached.

By the end of the war aircraft had generally standardized toward an open cockpit which contained an air speed indicator, an engine tachometer, an inclinometer (with some inherent error due to aircraft acceleration), an altimeter, and a compass (which also had errors due to aircraft vibration and an innate Northerly turning error).

Altimeters relied on changes in air pressure to provide a readout of the aircraft altitude. This indication of height arose from the Ideal Gas law which relates gas

Fig. 7.4 SE 5A restored World War I aircraft cockpit and instruments (copyright Dan Patterson used with permission)

density (ρ) to the pressure, temperature, and Ideal Gas constant of air. If we treat air as a fluid, then the change in the height of the air column (altitude) will be proportional to changes in pressure. This linear change in pressure equal to altitude assumes that the air temperature remains constant. If the air temperature were unchanged, then we would have an indication of altitude purely based on pressure differences between the ground and flight altitude. We know that air temperature varies with altitude plus season and scientists were investigating and publishing on this temporal temperature variation with altitude. Their work included seasonal variation for various regions in Europe and North America [6]. This temperature/altitude data allowed them to create a reasonably accurate altimeter based on the barometric pressure recorded on the plane coupled with the Ideal Gas equation.

Aircraft manufacturers and the pilots who flew their product both recognized the utility of having a compass in the cockpit. These devices were important for navigation, even though these early model aircraft rarely lost sight of the ground. However, the first magnetic compasses suffered large variation in direction, due to aircraft vibration and the much quicker change in direction from these airborne platforms vs. their traditional mounts (e.g., ground vehicles and ships). A more stable type of compass (gyro-compass) would not become available until after World War I.

The interwar years (1919–1938) were marked by large increases in airplane engine power, airframe size, and instrument performance. As previously noted, the Sopwith Camel biplane had a top speed of ~115 mph carrying only its pilot and

Fig. 7.5 Lindbergh's
"Spirit of St. Louis"
instrument panel

could reach 19,000 ft. altitude in 1917 [7]. A scant two decades later (1938), the single-wing Boeing Stratoliner could carry 33 passengers plus 5 crewmembers over 2300 miles in pressurized comfort (26,200 ft. service ceiling), while cruising at 220 mph [8].

However, even by Lindbergh's historic 1927 transatlantic flight, little had been added to the aircraft instrument panel. His plane incorporated an inductive compass (which failed during the flight), a conventional magnetic compass (located overhead to minimize metal interference, which meant that Lindbergh needed a mirror to read it), an oil pressure, and temperature gauge to monitor engine performance, along with the typical altimeter, inclinometer, and tachometer that had been primarily standardized during WWI. Figure 7.5 shows a photograph of the instrument panel from Spirit of St. Louis.

Like Lindbergh's record-setting airplane, commercial aircraft of this time used a similar set of instruments. The Ford Trimotor relied on nearly an identical suite of instruments like those used in Lindbergh's Ryan monoplane. The trimotor had additional oil temperature and pressure gauges, one for each engine. Figure 7.6 shows the layout of these instruments in the 2 seat Trimotor cockpit.

In addition to the instruments and their location, a more fundamental issue was being resolved around 1930. Prior to this time, standard design practice had the aircraft cockpit open to the airstream. But the increase in aircraft speed, typical flight altitude and a desired improvement in pilot comfort, dictated a closed cockpit.

Fig. 7.6 Ford trimotor cockpit ~1927

By the start of WWII most military platforms (fighter, bombers and reconnaissance) had enclosed cockpits. Enclosing the cockpit, while providing increasing pilot safety and comfort, removed the last vestige of direct input to the pilot from the airstream. Pilots continued to receive sensory input from the feedback on their control surfaces (prior to the advent of hydraulic assisted control surfaces) but they no longer could "feel" the wind and surmise information about their airplane based on air flowing over their head.

The last decade between the two major wars (1929–1938) saw an increase in the number and sophistication of instruments in the cockpit. This change arose from a desire to improve flight during night and bad weather, but also from natural progress in instrument quality and quantity.

American military aviator Jimmy Doolittle (later famous for the raid on Tokyo in 1942), long an advocate for the capabilities of military aviation, demonstrated the effectiveness of instrument flying in 1929. His airplane cockpit windscreen was covered, making visual cues impossible, but the aircraft had been modified with radio beacon receivers (to determine appropriate heading) and gyroscopes to generate an "artificial horizon" (to identify pitch up or down). He took off, made several turns, flew over the airfield, turned again, and landed his aircraft, without sight of the outside world [9].

The gyroscopic turn indicator started to become a reliable, and important, instrument at this time. It had become obvious to many (although a significant number of pilots resisted) that relying on the senses of the pilot could not provide a reliable gauge for aircraft attitude. The first gyroscopic attitude indicator relied on a sort of "spinning top" using an air-driven gyroscope mounted in the vertical axis. This axis

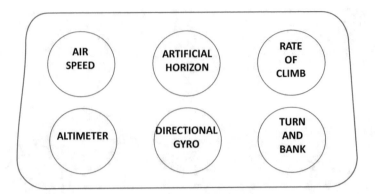

Fig. 7.7 Standard blind flying cockpit instrument locations [9]

would essentially remain vertical (due to gyroscopic force) as the plane maneuvered; therefore, the position of the gyroscope end on a hemispherical dome could show any deviation from level flight.

In addition to the artificial horizon indicator pioneered by Doolittle's flight, this time period saw the first rudimentary directional gyro vs. the tried, but frequently "not true" magnetic compass. Finally, organizations responsible for design and standardization (e.g., Royal Air Force) began advocating for standardization of instrument locations. Figure 7.7 shows the arrangement proposed by the RAF in 1937 for a cockpit capable of blind flying.

The interwar years saw the installation of numerous radio beacons for navigation within the USA. By the end of 1938, 6 were complete, but 159 more were in progress. By May 1941 (still pre US involvement in WWII), the Civil Aeronautics Administration (CAA, the major precursor to the FAA) began operating its first ultrahigh frequency (UHF) radio system for airline navigation, eventually reaching 35,000 miles of federal airways [10].

The Second World War saw a move to greater standardization in the cockpit layout (albeit hampered by the frequent increase in instrument capability), as well as a tremendous growth in instrumentation production to meet the enormous increase in quantity needed by the massive increase in military aircraft quantity, especially the growth in multi-engine aircraft that would fill out the bomber and transport inventory. These larger aircraft frequently had longer missions and increased navigation requirements. This time witnessed an increase in technological capabilities for the instrument suite as well.

Instrument flying became an accepted practice overcoming the previous mindset that "pilot senses" were paramount, supplanted by a willingness to believe and rely on the instruments within the aircraft. New instruments that arrived during this time period included new devices to determine the heading of the aircraft, which combined the best features of the magnetic compass with the directional gyros. In combination, many areas of the USA and Europe instituted ground-based radio -beacons, coupled with aircraft receivers. This duo increased pilot information, and greatly increased position knowledge. These radio beacons continued as aircraft navigation beacons for decades. These same ground-based radio beacons that began in the USA expanded to Europe and increased their proliferation primarily through the US Army Air Corps and our allies during World War II.

The ultrahigh frequency (UHF) radio beacon towers provided directional information, but also helped the airplanes maintain their position in corridors across the USA, and eventually throughout the world. While airplanes are not physically constrained to "highways" like automobiles, the control authorities (CAA now FAA) have long recognized the importance of guiding aircraft through designated "lanes" when flying. These designated travel "lanes" kept aircraft apart by spacing them horizontally based on their direction and velocity. The Commerce Departments Aeronautics Branch received the Collier Trophy for 1928, for the development of these air navigation aids. The Collier Trophy is an annual award for outstanding contributions to aeronautics or astronautics. By the end of the war, the CAA had included very high frequency (VHF) into their towers that included a VHF omnidirectional radio range that provided more accurate navigation information. A more accurate navigation cue came from the introduction of pulsed signals from 2 stations also during World War II. This system, the long-range navigation (LORAN) also supported maritime navigation. These radio beacons remained viable navigation aids through the 1960s when NASA and the FAA began using orbiting satellites to advance aircraft guidance. In addition, these "air highways" used vertical spacing to keep planes flying in opposite directions at least 1000 ft. apart.

The artificial horizon gyro unit inaugurated by Doolittle in 1929 retained that same appearance but achieved almost universal adoption. By the end of the war these devices had become adaptable to larger aircraft maneuvers including complete freedom in the roll and yaw axis, plus they could discern pitch angles ± 85°. The period during WWII saw the first introduction of heaters on airspeed indicators to prevent icing problems, a technique that has continued through the jet era. Figure 7.8 shows a rudimentary setup for this artificial horizon device which clearly lets a pilot know the orientation of his aircraft. Instrument designers have also been able to add additional functionality to these devices, including combining the output of the directional gyro, so that the aircraft heading is included in the artificial horizon (center instruments Fig. 7.8).

The aircraft designers attempted to provide logic to the haphazard method previously used to locate instruments within the cockpit. The single engine Supermarine Spitfire from 1943 is shown in Fig. 7.9. This cockpit demonstrates the engineers attempt to standardize instruments in the center view area, with additional switches and dials located outside this area. It is important to remember that the primary function of the Spitfire was to attack other aircraft, therefore its gunsight dominates the top-center of the view (Fig. 7.9).

The Douglas DC-3 commercial transport, which first flew in 1935, represented the culmination of prewar passenger aircraft technology and has been acknowledged as one of the greatest transport aircraft of all time. This twin-engine craft made airlines capable of profitability simply by flying passengers; they no longer needed to rely on government airmail subsidies. While it was a two-engine, piston/propeller aircraft, its cockpit represented many of the innovations that would be a

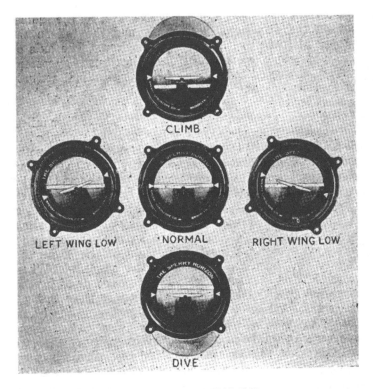

Fig. 7.8 Rudimentary artificial horizon instrument (~1930) [10]

Fig. 7.9 Supermarine
Spitfire Cockpit Circa
1943 [11]

Fig. 7.10 DC-3 cockpit, prototypical cockpit arrangement until the jet era [12]

hallmark of that aircraft control space, the cockpit. Figure 7.10 shows the inside of the DC-3 cockpit, a much more complicated affair vs. the Ford Trimotor. Engineers now had to cram dials and gauges onto the ceiling above the pilot and copilot to accommodate all of the added systems that became commonplace by the end of the propeller era.

After World War II, aircraft instrumentation displays did not change markedly, even with the introduction of the jet engine. This switch in engine type just meant more temperature and pressure measurements and their subsequent cockpit display. However, developments in electromechanical technology meant that mechanisms could now be fit into a remote location and electronic data sent to the cockpit for readout. This greatly reduced the number of pressurized lines entering the cockpit, thereby improving safety and reliability. This change also provided for a reduction in weight, as data transfer cables tend to be much lighter than metal tubing.

The lower center portion of Fig. 7.10 shows the throttle control levers. This central location allows both the pilot (left-hand seat) and copilot (right) to set the various engine thrust adjustments. These handles allow them to adjust propeller pitch ("bite" angle), engine fuel mix levers (to adjust as the airplane changes elevation and the mass of air varies), and the carburetor heaters to prevent icing in this vital fuel atomization step. Just above these throttle control levers are the four gauges related to engine performance. From left to right, they are engine RPM, engine manifold pressure (vacuum level), fuel pressure, and outside air temperature.

Figure 7.11 shows the view from the pilot's seat (left-hand side) in the cockpit. This view shows the standard yoke design that is still retained by Boeing commercial aircraft and the major gauges, dials and switches directly in front of the pilot.

Fig. 7.11 Pilot view of DC-3 instrument panel [12]

Fig. 7.12 Boeing 707 Cockpit 1962 [13]

The top center gauge is the artificial horizon, with the airspeed indicator to the left and the altimeter to the right. A compass heading is seen below the artificial horizon and the secondary altimeter is below the primary altimeter on the right.

All of the instruments, gauges, switches and dials have now been increased in number and complexity as aircraft entered latter half of the twentieth century. Figure 7.12 shows the cockpit a Boeing 707 that was used as Air Force One. The US Air Force designated this model a VC-137, Boeing delivered this aircraft to the

Fig. 7.13 Modern flight deck with "glass" cockpit and sidestick controller

US Government in 1962 for use by the President and other high-ranking government officials [14]. Only when the president is on board does it receive the call sign "Air Force One." This photo shows the complexity already found in cockpits during the jet era, and the four engines on this aircraft just added to the instrumentation and gauges to be monitored. In fact, not seen in the photograph is a flight engineer station, located behind the copilot seat, which would reveal a similar array of dials, gauges, and radio equipment.

Figure 7.13 shows the most recent incarnation of a modern airliner. This view demonstrates the increase in digital instrumentation, and the use of flat screen displays (sometimes referred to as the "glass" cockpit) that can be used to show a myriad of sensors and system operating conditions. While the sidestick controller has been adopted by Airbus and many military aircraft suppliers, Boeing still retains its central yoke and control wheel setup. Finally, many navigation (GPS) communication and radar improvements have been incorporated into the cockpit. However, the fundamental method for aircraft velocity, ice detection, and angle of attack has not changed for decades. Aircraft still use a pitot probe external to the fuselage to measure airplane velocity. Angle of attack sensors use either a trailing vane or a slotted vane to discern the aircrafts pitching angle compared to the airstream. Ice-detecting sensors can rely on a cylinder exposed to the airflow. This cylinder is made to vibrate and the added mass of ice on the cylinder will change the resonant frequency. This change in frequency can be calibrated to the amount of ice buildup.

The fundamental sensors (angle of attack, air velocity) and even the artificial horizon have not changed dramatically in the change to jet propulsion, or even the new digital cockpit transformation. Figure 7.14 shows a pitot probe used for measuring airspeed, with an imbedded electric heater, to prevent ice buildup from generating a false reading.

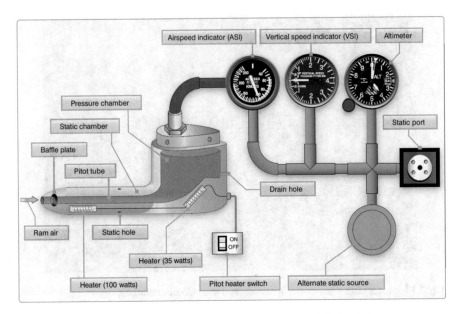

Fig. 7.14 Modern-day pitot probe for aircraft airspeed with embedded electric heater

Navigation has seen a dramatic improvement since the early airmail days when large fires were sometimes used to provide navigation markers or identify landing sites at night. As previously mentioned, the adoption of radio beacons provided a large increase in aircraft safety and improved navigation. Researchers at the Johns Hopkins University, Applied Physics Laboratory (APL) discovered that the position of the Soviet Sputnik satellite could be determined by careful measurement of the Doppler shift in its continuous-wave transmitter. They then surmised that your earth location could be determined by observing the Doppler shift of a stable transmitter from a satellite, the first suggestion of the Global Positioning System (GPS) [15].

The genesis of the GP System arose from military requirements. The Navy wanted a way to accurately locate the position of their ballistic missile submarines prior to launch, to improve the accuracy of their targeting. These initial GPS studies recommended a system with three features in common. First, the system had to accurately determine the position of the satellite. Second, it had to use radio waves that could penetrate the ionosphere and clouds (all weather ability). Finally, it had to measure the signals relative time of arrival from time-synchronized satellites. The satellites relied on atomic clocks to synchronize their time to a common standard. The system was demonstrated via balloon transmitters in the early 1970s. However, it was not until the 1990s that the full operational capability of the system was declared (end of 1994). Despite, this the system was successfully demonstrated by the military during the first Gulf War (Aug 1990–Feb 1991).

While the GP system had military roots, it has now grown into a worldwide civilian infrastructure that supports the aerospace, maritime, and land-based enterprises as docile as farming.

References

1. Aviation Maintenance Technician Handbook, Vol. 2, Ch 10, FAA-H-8083-31, 2012
2. https://en.wikipedia.org/wiki/Flight_altitude_record, retrieved 7/8/17
3. Smithsonian National Air and Space Museum
4. Seventy years of Flight Instruments and displays, The Aeronautical Journal, R A Chorley, Volume 80, Number 788, 1976, pp 323–342
5. http://www.century-of-flight.net/Aviation%20history/airplane%20at%20war/world%20war%20one%20aircraft%20statistics.htm accessed July 19, 2018
6. Aircraft Instruments, Stewart, C. J., John Wiley & Sons, NY 1930
7. https://en.wikipedia.org/wiki/Sopwith_Camel retrieved 7/28/18
8. http://www.boeing.com/history/products/model-307-stratoliner.page retrieved 7/29/18
9. Flying Blind, Matt Mahoney, January 2, 2013 MIT Technology Review
10. https://www.centennialofflight.net/essay/Government_Role/navigation/POL13.htm retrieved 1/3/19
11. Flying an Airplane in Fog, Doolittle, J. H., SAE Journal, Vol. 26, Issue 3, Mar 1930, p. 318
12. Photo courtesy of Dan Patterson, used by permission
13. http://www.maam.org/ retrieved 10/7/18, used by permission
14. https://www.nationalmuseum.af.mil/Visit/Museum-Exhibits/Fact-Sheets/Display/Article/195807/boeing-vc-137c-sam-26000/ retrieved 10/7/18
15. The Global Positioning System, Ivan Getting, IEEE Spectrum, December 1993

Chapter 8
Anti-ice and Deice Systems for Wings, Nacelles, and Instruments

The need for anti-ice or deicing systems did not show up during the first one and one-half decades of the aviation industry. Pilots during the first decade and into the First World War relied on visual cues for navigation. This reliance on ground-based navigation aids and a limited service ceiling meant that pilots would only rarely, and almost never, intentionally encounter potential icing conditions in flight.

The boon to aviation that airmail brought in the 1920s also delivered pressure to meet delivery schedules. The need to deliver the mail, coupled with the inherent altitude requirement of the Chicago to the West coast flight connection and inherent vagaries of weather throughout this service territory, insured that icing conditions would now start to impact aviation.

It would take numerous losses of men and planes for the problem to become significant enough to warrant a thorough investigation. Congress established NACA (National Advisory Committee for Aeronautics) in 1915 to supervise and direct problems in flight for the US aviation industry. Initially NACA was investigating potential improvements in flight characteristics including propellers, engines, and wings. NACA eventually became incorporated into NASA in 1958. However, by 1928 NACA had begun investigating the problems of icing in flight.

The problem with investigating icing formation in 1928 was the lack of understanding for the conditions that could produce ice on aeronautical surfaces. The limited refrigeration capabilities of the time, Freon the first safe refrigerant did not become available until 1928. Finally, their lack of ability to generate extremely small water droplets would also make it difficult to duplicate in-flight icing conditions in an engineering lab [1].

Ice and snow can accumulate and affect lift and control while an aircraft is on the ground. Airports routinely provide deicing services to prevent this accumulation from impacting takeoff, but the crash of Air Florida Flight 90, a Boeing model 737, into the Potomac river on January 13,1982, highlighted the importance of this function [2]. However, we will focus on changes to the aircraft that occur while it is flying through icing conditions. To have ice develop in flight, liquid water must be present in the atmosphere. However, as previously noted, the atmosphere's

© Springer Nature Switzerland AG 2020
T. Filburn, *Commercial Aviation in the Jet Era and the Systems that Make it Possible*, https://doi.org/10.1007/978-3-030-20111-1_8

temperature continues to drop with increasing altitude, with temperatures easily dropping to −4 F at an altitude of 20,000 ft. The ability for the atmosphere to hold water is dependent on its temperature. At altitudes above 22,000 ft. the temperature has declined sufficiently that the liquid water content (LWC) of the air is sufficiently low that icing conditions rarely occur above that altitude.

We are familiar with water forming ice and snow as its temperature drops below freezing (0 C); however this phase change (liquid to solid) does not automatically occur with dropping temperatures. Water can remain in its liquid form at temperatures well below its normal freezing point. It generally requires a nucleation site for ice to form at temperatures above −40 C, dirt or ice in the atmosphere can generate ice crystals. In the absence of initiation sites, these supercooled water droplets can form ice crystals when they impinge on a surface (nacelle, wing, horizontal stabilizer). The formation of these ice crystals on the leading edge of these surfaces can provide a bonding surface for additional ice buildup and accretion. Many early aviators learned of this peril when their airplane lost lift due to ice buildup on their wings, or loss of control due to ice limiting control surface movement or altered airflow. These early airplanes had limited thrust capability and the ice accumulation also increased drag. The drag could become high enough that the low engine power could not keep the plane above its stall speed, which could also jeopardize flight. Equally troublesome were ice forming on instrument surfaces (pitot probes for measuring velocity) or propeller blades, both incidents could and did bring planes down. Finally, ice buildup can lead to aircraft imbalance or the additional weight of the ice can make it difficult to control the aircraft. Rotary winged craft share many of these problems associated with ice buildup, but we will confine our discussion to fixed wing aircraft.

Ice accretes on plane surfaces initially at the stagnation points. Those are the leading surfaces that exactly divide the flow around that surface. Examples of stagnation points include the wing leading edge, as well as rudder, horizontal stabilizer, and instrument leading surfaces. Figure 8.1 shows a wing stagnation point, the most likely initiation site for ice buildup. Propellers and the spinners that provide aerodynamic cover over the central propeller region also need ice protection, as they also have stagnation points that regularly receive ice buildup while flying in icing conditions. The aerodynamic property of the propeller that provide thrust can be greatly altered by ice buildup, and the spinner, designed to reduce drag, can suffer a large increase in drag with sufficient ice buildup.

By February and March of 1928 both the Army Air Corps and the Navy's Bureau of Aeronautics had approached NACA with a request to investigate the problem of airplane icing. This investigation should determine the conditions that could lead to icing, find preventative measures, and develop instruments to tell pilots when ice formations could occur [3]. This same year, 1928, the Langley lab began operating

Fig. 8.1 Ice buildup at wing stagnation point [3]

a small diameter (6 in.) wind tunnel that could operate in icing conditions. Unfortunately, its small diameter, and the inability of engineers to duplicate the very small drop size found in nature, made its ability to generate breakthrough research difficult at first.

As early as 1930 researchers had investigated a host of materials to prevent ice formation on wings, propellers, and control surfaces, with no long-term success (an investigation that continues today). They looked at organic materials, hydrophobic substances, and coatings that repelled water (e.g., oils and greases). While the coatings provided short-term relief, they regularly sloughed off the wings, supports, and control surfaces. This meant a system would need to be employed to routinely readminister these coatings along with the parasitic weight of the replacement coating, which greatly detracted from their utility. Early investigators discounted the ability of high temp engine exhaust to be used for deicing because they thought the entire wing needed this capability. Obviously, engine heat would not be employed on the very early wing materials that included wood and canvas treated with aircraft dope to provide a rigid surface. Their initial analysis indicated insufficient heat energy in the engine exhaust to keep the entire wing surface above the freezing point. However, these researchers did indicate the possibility of using inflated and deflated bladders to shed ice, a system still in use today [4]. Figure 8.2 below shows the location of these boots at the leading edges of the wings, struts, radio mast, and tail control surfaces. These boots are still in use on general aviation and turboprop commuter aircraft, in the exact same location (minus the struts which have been deleted from modern aircraft). Two different types of devices can be employed to prevent ice accumulation, deicing and anti-icing systems. Deicing systems allow a small amount of ice to accumulate and then periodically shed this ice via mechanical or

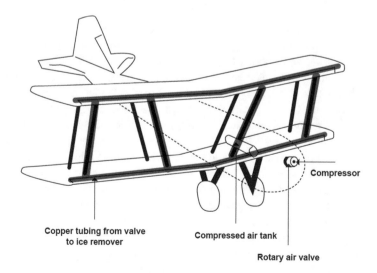

Fig. 8.2 Locations for inflatable rubber boots on early biplanes [4]

thermal means. Anti-ice systems prevent any ice from accumulating by evaporation or allowing liquid water to run off and potentially freeze on noncritical surfaces [5].

BF Goodrich, led by William Geer, were instrumental in the early deicing boot designs and continued to refine the inflatable boot concept. This idea, which seemed to have strong merit, after wind tunnel research indicated that clearing the stagnation point at the leading edge of the wing would allow the wind stream to strip the remaining wing surfaces of any accreted ice. This idea was installed on an Airmail plane in 1930 and then a Lockheed Vega (*Miss Silvertown*) with the boots put on the wings, struts, and tail surfaces. An air compressor automatically provided air to the boots. The New York Times announced (prematurely) "victory" over "one of aviation's most dangerous enemies." Airlines quickly lined up to adopt these devices. United Airlines ordered them for their Boeing 247s in the summer of 1933, while TWA installed them on their DC-2s and DC-3s [1] beginning in 1934. For TWA the cost of the deicing system (compressor, valves and boots) was about $65,000, and added nearly 180 pounds to the aircraft. But both financial and weight penalties seemed small compared to the hoped for improvement in safety provided by the system.

The key to the success from these inflatable boots is their ability to change the shape of the leading edge and initiate cracks within the ice layer. Due to the low ductility (inability to move without cracking) of the ice layer, the boot inflation moves the rigid ice layer, cracks it, and the high air velocity sweeps it away. Figure 8.3 below shows the concept for this inflated and deflated boot arrangement on both a wing and strut.

On March 26, 1937, all the optimism that surrounded the new pneumatic deicing system evaporated when a TWA DC-2 crashed while attempting to land at Pittsburgh's airport. The impact caused the death of all 13 passengers and 3 crew members but did not result in any fire. First responders to the accident scene reported seeing ice on the wings and control surfaces.

The loss of the Pittsburgh bound TWA flight restarted NACA's investigation into icing conditions and methods to prevent ice accumulation. The Langley laboratory built a larger scale wind tunnel to study icing, while in-parallel other researchers procured aircraft as flying test beds to investigate icing conditions and methods to mitigate its impact. The larger icing tunnel at Langley found itself being used for other purposes, while researchers used several different test aircraft to investigate the potential of using engine exhaust as a means to eliminate ice formation on the leading edges of wings. By 1940 NACA was convinced that the engine exhaust gas could be used to prevent ice formation at the wing leading edge (the most critical area). This testing and analysis indicated that sufficient energy was present in the exhaust gas to cover the wing leading edge in sufficient length to eliminate ice as a significant impediment to lift generated by the wing [6]. Engineers and material scientists worried that directly impinging the exhaust gas into the wing box area might overheat or increase corrosion of the now predominantly aluminum structure. Therefore, NACA engineers explored a heat exchanger in two different configurations, to extract the thermal energy from the exhaust and use warmed air to prevent ice accumulation on the leading edge of the wing. Figure 8.4 shows these two

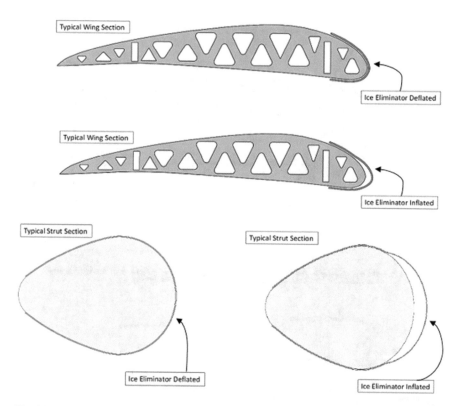

Fig. 8.3 Ice boots showing inflated and deflated configuration [4]

configurations, one in which the exhaust gas runs near the leading edge in a tube, and the second where the exhaust gas heats an air stream which then circulates inside the wing box.

While NACA had demonstrated the efficacy of engine exhaust as a wing deicing method, many World War II aircraft stayed with the pneumatic boot deicing system, including well-known Boeing platforms like the B-17 and B-29 (although they were removed during latter stages of the war from the B-29 when it was used in lower altitude missions).

Smaller general aviation (GA) aircraft, along with smaller turboprop commuter aircraft, still operate with pneumatic deicing boots today. These boots cover the wing, rudder, and horizontal stabilizer, at the leading edge of each listed surface. They are regularly inflated and deflated during flight while in known or suspected icing conditions. By breaking up the ice at the leading edge, they allow the aerodynamic airflow to break up the ice downstream of the leading edge, because of the sudden dimensional change caused by the loss of ice at the leading edge. Larger commercial aircraft (e.g., longer range, larger load jet aircraft) rely on different technologies vs. the pneumatic boots. While the boots represent a lower initial cost system, they add drag, especially costly for aircraft flying at $M = 0.85$ on routes

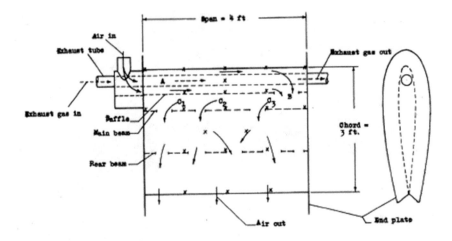

Figure 3.- Set-up of wing heated by exhaust tube, showing path of exhaust gas and air through the model and the location of skin-temperature thermocouples, x.

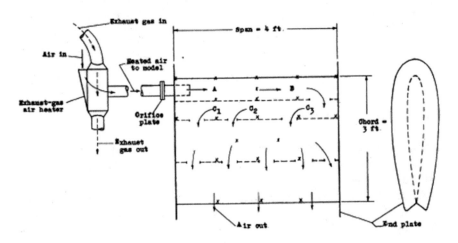

Fig. 8.4 Air heated via engine exhaust NACA test apparatus

covering hundreds to thousands of miles, and they have limited life, needing to be replaced after several thousand hours of operation.

The propeller blades on the same GA and turboprop aircraft also need deicing to maintain their ability to generate thrust while flying through icing conditions. Larger turboprop commuter aircraft tend to use electric heating elements at the base of the propeller along with heaters on the spinner section to prevent ice buildup. Eliminating ice at the root of the blade allows the large centrifugal forces to remove any remaining ice from the outer regions of the blade. A slip ring assembly allows for continuous electricity supply to the propeller while it is rotating in flight.

Larger commercial aircraft like the Boeing 737 or the Airbus A320 will rely on engine bleed air to prevent ice accumulation on critical wing surfaces. Engine bleed is air that is taken from some point in the compression process of the engine. Thermodynamic laws dictate that a gas (air) temperature will increase during any compression process. In modern axial flow compressors used by turbofan engines this heat of compression can easily achieve compressor exit temperatures over 1000 F even at their cruising altitude where inlet gas temperatures are -50F. This 1000 F temperature is before any fuel has been combusted, meaning that combustor exit (and therefore turbine inlet) temperatures can routinely rise 800 F or more above these compressor exhaust temperatures. As these are multistage compression devices, intermediate points can be selected for lower gas temperatures.

The same large commercial aircraft models made by Airbus and Boeing will duct some small fraction of engine airflow into the wings for heating the leading edge. These original equipment makers (OEMs) have both settled on designs that use the engines slung on pylons below the wings. With this geometry, it is very convenient to take hot "bleed" air from the compressor section of the engines to heat the leading edge of the wing. This deicing function typically cycles on/off to limit the impact on fuel burn. Every pound of hot compressed air that is circulated inside the wing is a parasitic loss on engine thrust and efficiency. Figure 8.5 shows how this air is generally circulated through the leading edge and then is allowed to escape into the free stream around the wing.

Surprisingly, not all of the wing leading edge surface needs to be heated for this deicing measure. Figure 8.6 shows the range in wing span covered by the thermal engine bleed deicing system for several Boeing model airplanes. The trend for reduced percent span also follows the general trend of the date when that model was

Fig. 8.5 Wing leading edge bleed air deicing heating flow [7]

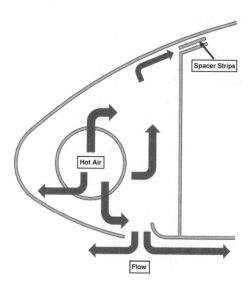

Wing Leading Edge Bleed Air Deicing Heating Flow

Spacer Strips

Hot Air

Flow

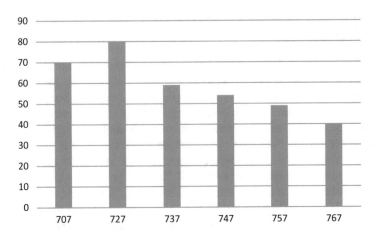

Fig. 8.6 Wing ice protection, percent span for select Boeing airplanes [7]

introduced. The 707 (first flight December 1957), the first commercial jet liner from Boeing, relied on 80% span coverage. Whereas the 767, which first flew in September 1981, had only 40% of its wing span capable of being deiced in-flight. This trend demonstrated the conservatism of the early designers along with the design changes allowing more ice to be shed by adjacent heated regions.

For the predominant design choice of Airbus and Boeing (engines on underwing pylons), the empennage (tail assembly) does not have the luxury of close proximity to the engines which the wings enjoy. Therefore, electric heating elements, pneumatically inflated boots, and no ice protection have been implemented on these tail surfaces for aircraft from these manufacturers. For the smaller percentage of large commercial aircraft with tail-mounted engines, this will shorten the heating distance and allow engine bleed to deice the empennage surfaces. For the case of no ice protection, airplane designers have simply increased the size of the rudder and horizontal stabilizer to account for the potential for ice buildup and to ensure that sufficient yaw and pitch control is still achievable.

As previously mentioned, pneumatically inflated boots have been used for ice protection since the 1930s. While they were the first active in-flight deicing system, they are still retained on several classes of airplanes, due to their simplicity, low weight, and low power draw. Systems that bleed compressed engine air to thermally melt ice provide a significant decrement to engine efficiency. However, rubber boots glued on to the leading surfaces can by cycled between inflated and deflated positions regularly to provide ice protection.

The rubber boots shown in Fig. 8.7 tend to be used on smaller commercial aircraft like the turboprop commuter size, as well as on general aviation airplanes. The boots are frequently installed on these types of aircraft because they provide deicing protection at a significantly lower energy penalty vs. engine bleed or electric heater blankets. Larger aircraft with larger engines and larger electric generators will avoid the boots because they require more frequent maintenance and because of the

Fig. 8.7 Cross section of pneumatic deicing boot deflated (top) and inflated (bottom) [8]

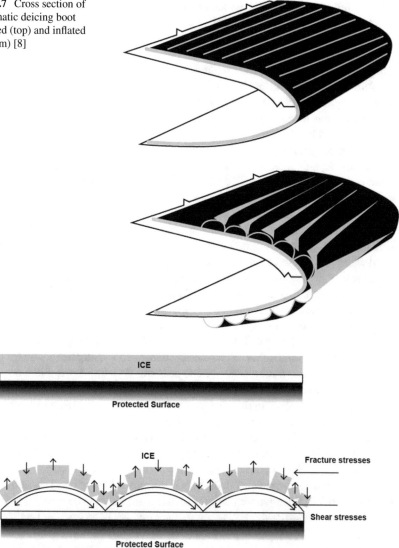

Fig. 8.8 Effects of boot inflation on ice accretion [9]

increased drag imposed by this design. The increased drag becomes a severe impediment at the higher cruising speeds of the Airbus and Boeing aircraft. With the pneumatic boots installed on the wing leading edge, they are very prone to damage from bird strike, natural objects (hail and rain can both erode the rubber), and UV degradation from sunlight.

The rubber boot deicing systems use a brief inflation time and longer deflation time to achieve their deicing function. Figure 8.8 shows how the 1″–2″ diameter tubes are inflated to break up the ice buildup. Aerodynamic forces help remove the ice from the following surfaces due to the large discontinuity generated by the leading ice of the ice breaking and sloughing off the aircraft. The tubes receive air pressurized between 25–50 psig, for 5–10 s. After that inflation period, a vacuum is pulled inside the tubes to quickly deflate them and insure that they are fully retracted to minimize the drag imposed on the airplane. This inflation/deflation cycle is repeated approximately once per minute.

Turbofan and turboprop aircraft need ice protection for their nacelles (engine housings) in addition to the deicing needed for their wings and control surfaces. This nacelle ice protection is meant to keep large ice sections from being ingested into the engine, and potentially damaging the high-speed fan and compressor blades or the stationary guide vanes. For this reason, a periodic deicing system like the pneumatic boots is generally not favored for these engine/nacelle locations. Most nacelles use hot engine bleed air due to their close proximity to the engine. This air is circulated through the nose lip of the nacelle in a very similar geometry to the wing leading edge hot air systems, in order to prevent ice from accumulating. The liquid water is collected or allowed to run through the engine.

As previously described, electric heaters can be employed when bleed air cannot be easily circulated to the surface requiring ice protection. These electric heater mats generally require high energy and energy density due to the large convective cooling effect of the high-speed air stream, and the large temperature difference required, easily needing >10 kW of power for several wing sections. These deicing systems tend to be some of the highest electric loads on the electric generator geared off the engine. These surfaces must be kept above 0 C, but the ambient air can be −50 C, with a relative velocity of 550 mph. In addition, as a minimum these heaters need to supply the energy to melt the ice (sometimes evaporate it), which will create even higher power draw on the aircraft electric grid.

A final section of the aircraft that requires deicing is the various instrument nozzles located near the cockpit and the engine to track the performance of both the aircraft and the engine. As introduced in Chap. 7, several instrument ports are located near the cockpit to provide vital flight and aircraft information. The pitot probes which extend into the airstream from a small fairing need to be ice-free to provide aircraft velocity measurements. Therefore, they frequently rely on continuous heater power as long as the aircraft is powered (even before takeoff). The total air temperature probes along with angle of attack instruments also need to be ice-free to perform their function and rely on electric resistance heating. All three instruments can typically be found near the front of the fuselage, many with warning marking due to the high temperature they may reach before the aircraft becomes airborne.

References

1. We Freeze to Please, NASA SP-2002-4226, Leary, W.
2. FAA Lessons Learned, http://lessonslearned.faa.gov/ll_main.cfm?TabID=3&LLID=2&LLTyp eID=0, retrieved 3/23/2018

3. Moffett to Lewis, 12 March 1928, File RA 247, Langley Library
4. The Prevention of the Ice Hazard on Airplanes, Goer, William, Scott, Merit, NACA Technical Note 345, July 1930.
5. A handbook method for the estimation of Power Requirements for Electrical De-icing systems. DLRK, Hamburg, 31. August-02. September, 2010, Document ID 161191
6. A Flight Investigation of Exhaust-Heat De-Icing, Rodert, Lewis, Jones, Alan, NACA Technical Note 783, November 1940
7. Report of the FAA International Conference on Airplane Ground Deicing, Reston VA, May 28–29, 1992, FS-92-1
8. Aviation Maintenance Technician Handbook – Airframe Vol.2 (FAA-H-8083-31) Chapter 15, Ice and Rain Protection
9. Giving Ice the Boot, Understanding Pneumatic De-Icing, Technical Bulletin 101 Rev. D, Goodrichdeicing.com

Part II
Sub-system Accidents

Chapter 9
Loss of Flight Controls, United Flight 232

Early Type Accidents

The DC-10 was McDonnel Douglas answer to the request for larger aircraft from several airlines during the 1960s. By 1968 Douglas Aircraft Corporation, at one time the leading producer of airliners, had been bought out by the McDonnel Aircraft, demonstrating the vagaries of the airplane industry. The DC-10, a three-engine plane, could accommodate up to 380 passengers and represented the second introduction into the wide-body market. The Boeing 747, the most successful type in the wide-body field, was the first wide-body aircraft introduced into the market, followed by the DC-10 and then Lockheed's L-1011 model representing all of the US entries into this market.

The DC-10 like Lockheed's version was a three-engine airplane. The DC-10 airplane was slightly smaller aircraft than the Boeing 747 model with a smaller passenger load. The 747 could weigh over 700,000 pounds, while the DC-10 would be 70% that size weighing around 440,000 pounds. While trying to tap into the wide-body market, it would also be cheaper to purchase, operate with 1 fewer engine, and had a smaller airframe vs. the 747. While over 1,500,747 models would be produced by Boeing, McDonnell Douglas would only build about 450 of the DC-10 (including 60 for the military as the KC10 tanker variant).

The DC-10 placed two of its engines on underwing pylons, a very typical location for all vendors in the industry from the inception of the jet engine (note Germany's WWII fighter, the ME262). The third engine would be housed in a nacelle within the tail structure. Figure 9.1 shows a diagram of the DC-10 with its nominal dimensions and 3 engine placements. Engine 1 was placed on the left wing, Engine 2 was housed in the tail, and Engine 3 was placed under the right wing. All of the early models would operate with GE model CF6 turbofan engines, which could produce about 50,000 lbf of thrust each. The −40 model, the final transport variant of the DC-10, would also fly with P&W JT9D turbofan engines, which each could produce about 56,000 lbf of thrust.

© Springer Nature Switzerland AG 2020
T. Filburn, *Commercial Aviation in the Jet Era and the Systems that Make it Possible*, https://doi.org/10.1007/978-3-030-20111-1_9

Fig. 9.1 McDonnel
Douglas DC-10, three-
engine passenger aircraft,
with engine location

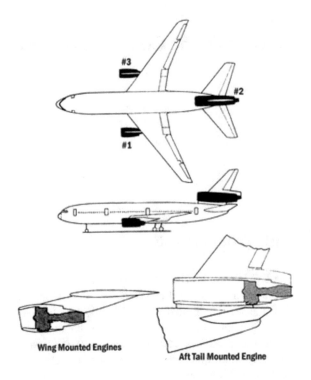

Wing Mounted Engines

Aft Tail Mounted Engine

   The DC-10 had a number of early accidents, several of which were attributed to
aircraft design flaws, while others were attributed to subsystem design or fabrication problems. Multiple accidents involved the rear cargo hatch, which originally
had a faulty latch and indicator assembly. The cockpit could receive a faulty indication that the hatch was fully closed and latched, when in fact the door was not properly latched. The rear cargo door opened outward (vs. the traditional passenger
compartment doors that must first swing inside the pressurized aircraft before opening). This outward opening mechanism provided improved access to the cargo hold,
and easier movement for ramp personnel. However, it also meant that a single failure could allow the door to depressurize the cargo compartment and then the entire
airplane. Figure 9.2 below shows the location of the left, aft cargo door and its original, suspect latching mechanism.

   American Airlines Flight 96, a DC-10 from Los Angeles to La Guardia, with
planned intermediate stops in Detroit and Buffalo, suffered a depressurization accident. On the afternoon of June 12, 1972, the flight landed in Detroit and after disem-
barking some passengers and adding others, the plane prepared for its next stop,
Buffalo NY. Ramp personnel removed and added cargo to the aft cargo hold, but had
difficulty closing the aft cargo door. The ramp attendant working to prepare the
aircraft for departure consulted an airline mechanic about his difficulty with the
door, but received an ok to proceed. The indicating light in the cockpit that indicated
if the door was not latched properly, never illuminated, leading the flight crew to

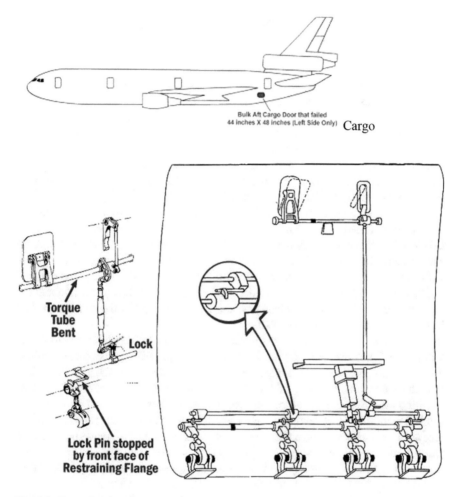

Fig. 9.2 Cargo door location and latch mechanism

believe they were safe to proceed. As the aircraft took off and climbed away from Detroit, both the cockpit personnel, passengers and flight attendants heard an audible "bang" as the plane climbed to almost 12,000 ft. At this altitude the pressure difference inside to outside the aircraft would be merely 1.4 psia. However, this small (1/10th atmosphere) pressure differential could produce nearly 3000 lbf of force acting to push open the door, because of the greater than 2000 in [1] door area.

The rapid depressurization, acting through the lower cargo hold, buckled the main passenger floor because there was no easy way to relieve pressure between the now ambient (12,000 ft. altitude) cargo hold and the higher pressure passenger compartment. The floor buckled, and but for the low loading on the floor (only 56 passengers out of a capacity of over 250), it might have produced more significant floor deformation and damage. Control cables for the rudder and horizontal stabilizer ran through the buckled floor, making it difficult time for the pilots to control the aircraft.

The pilot was able to return the aircraft to Detroit, where he safely landed the plane without any loss of life or additional damage to the aircraft [2]. This accident would serve as an eerie premonition of the Sioux City Accident in 1989, in which all aircraft flight control, through the tail surfaces, were lost.

Unfortunately for the flying public, the response to the American Airlines accident that departed and then had to return to Detroit did not carry the urgency demanded by this design flaw. McDonnel Douglas issued a service bulletin to modify the door and latch mechanism. Unfortunately, the FAA did not issue an Airworthiness Directive that would have grounded DC-10s until design changes were incorporated. Airlines were required to inspect their cargo door mechanism and had 300 h to implement the relatively simple changes McDonnel-Douglas suggested in their post-accident service bulletin [1].

However, the loss of Turkish Airlines flight 981 would resurrect the problem of the DC-10s aft cargo door. This DC-10 was leaving Paris' Orly airport bound for London. The plane was fully loaded with 346 passengers and crew when it took off early on the afternoon of March 3, 1974, only 21 months after the cargo door depressurization accident over Detroit. About 5 min after taking off, as the plane climbed to 12,000 ft. the cockpit voice recorder captured noises of decompression and the copilots voice stating, "the fuselage has burst." The plane pitched down and crashed in a forest 25 miles northeast of Paris.

Once again, the same aft cargo door had opened due to inadequate latching and a buildup of pressure from the decrease in external pressure as the plane climbed. This time the fully loaded passenger compartment produced a much larger buckling into the cargo compartment of the floor supporting the passenger seating. This buckling was so severe that all control over the tail section (horizontal stabilizer and rudder) were lost, along with the ability to adjust the thrust on the tail-mounted #2 engine. The buckling was initiated by the sudden pressure difference between the cargo compartment and the passenger compartment. The rapid depressurization and force of the floor buckling ejected two rows of passenger seats along which were found 8 miles from the crash site along with the cargo door. Figure 9.3 shows the larger floor load from the higher passenger count with the Turkish Airlines during its depressurization accident. That higher floor load led to a complete loss of control for all of the tail flight control surfaces (rudder, horizontal stabilizers). This Paris accident provided an eerie premonition for the challenges that the United Airlines Flight 232 crew would have to handle.

This time the FAA generated several Air Worthiness Directives for the DC-10 which mandated improvements to the cargo door latching mechanism. Additional Air Worthiness Directives expanded to include additional wide-body aircraft including the DC-10, Boeing 747, Lockheed L-1011, and Airbus A300. These final directives required the manufacturers and airlines to reinforce the floor between the cargo and passenger compartment, such that a sudden pressure loss in one region would not produce a collapse of the floor [3].

The depressurization accident over Windsor Ontario, and the complete loss of the airplane with all on-board while leaving Paris, generated very strong negative public opinion for the DC-10 aircraft. It would get worse with the next accident in 1979.

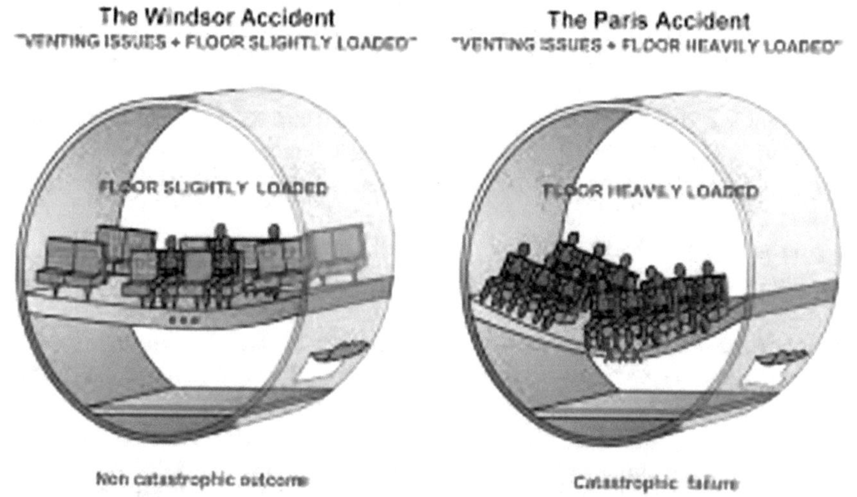

The Windsor Accident
"VENTING ISSUES + FLOOR SLIGHTLY LOADED"

The Paris Accident
"VENTING ISSUES + FLOOR HEAVILY LOADED"

FLOOR SLIGHTLY LOADED

FLOOR HEAVILY LOADED

Non catastrophic outcome

Catastrophic failure

Fig. 9.3 Floor buckling from partially loaded vs. fully loaded passenger compartment

The next accident was an American Airlines DC-10 (flight 191) leaving Chicago's O'Hare airport was a complete loss. The flight was accelerating down the runway in preparation for takeoff, when the left engine (#1) departed from the plane in a dramatic fashion. The engine broke away from its support pylon and rotated over the left wing. This happened exactly at the velocity where the pilots begin rotating the nose to leave the ground.

The airplane is designed to be able to continue into flight with the loss of the thrust from one engine, the problem with this DC-10 accident was the way the engine left the aircraft. It severed control cables and hydraulic lines servicing primary and secondary flight control surfaces. The aircraft also lost the electric power generated by the #1 engine, which included several critical features like stall warning and the Captain's flight instruments. This plane crashed barely 30 s after taking off. All 271 persons on the plane, as well 2 people on the ground were killed in the crash. The immediate cause was the retraction of the slats outboard of the engine on the left wing, which reduced the lift on the wing to point of stalling. The slats retracted due to the lack of hydraulic pressure to keep them in their takeoff deployed position. The hydraulic pressure failed because of the separation of the engine and pylon from the wing. With the loss of lift on the left wing, the airplane rolled to its left and crashed. The root cause of this Chicago flight was not due to airplane design. Instead, the airline performed maintenance on the wing and pylon in a nonstandard fashion, unapproved by the McDonnel Douglas. This maintenance procedure generated an unseen crack in the pylon, which failed during the takeoff event in Chicago [4].

United Airlines Flight 232 (UAL 232)

United Flight 232 was a McDonnel Douglas DC-10 flight from Denver to Philadelphia with a planned stop at Chicago's O'Hare airport. The flight left at 13:09 pm (MDT) on July 19, 1989, from Denver's Stapleton airport with 285 passengers and a crew of 11 on board. The -10 model of the DC-10 was the initial configuration of the airplane.

No abnormalities had been reported by the flight crew as they reached their cruising altitude of 37,000 ft. About 1 h and 7 min into the flight both the flight crew and passengers heard a loud bang. Several of the flight attendants and some of the passengers in the rear of the airplane thought a bomb had exploded in the plane [5]. The explosion produced an immediate shuddering and continual vibration in the aircraft. At the time of the explosion, the first officer was flying the aircraft. The cockpit crew noted the failure of the #2 engine (tail mounted) through a scan of the instruments and began reading and acting on the engine shutdown checklist. While going through this checklist, the airplane began a right descending turn, and the first officer alerted the pilot that he could not control the airplane. While continuing to go through the #2 engine shutdown procedure, the flight crew noted that all hydraulic systems were trending to zero pressure. This was indicative of a complete loss of all three hydraulic systems and therefore the inability to move any control surfaces (primary or secondary). The captain took control of the airplane but confirmed that the plane did not respond to control column/yoke inputs. The captain began to reduce thrust in the #1 (left) engine which produced a wings level attitude. This asymmetric thrust control was not part of any emergency procedure, as the aircraft designers and regulators thought that the loss of all hydraulic control was not a credible failure. This supposition was based on the multiple redundancies that were designed into the systems, the ability for any of the 3 engines to supply power to the hydraulic system and finally an emergency air-driven generator that could also be deployed to power these control lines.

As previously described, hydraulic systems had been introduced into airplane control schemes, before World War II. However, their popularity greatly increased during the war as military airplanes grew in size and speed. Both of these growth trends meant that human-powered methods to move airplane control surfaces would be inadequate. Airplane designers relied on hydraulic systems of increasing pressure to generate sufficient forces to move the control surfaces.

McDonnel Douglas continued this reliance on high-pressure hydraulic systems, in a similar fashion to the other large transport aircraft builders. In order to provide for crew and passenger safety, the company used three independent hydraulic systems, any one of which could generate sufficient forces to move the DC-10 primary flight control surfaces, secondary flight control (flaps and slats for takeoff and landing), retract the landing gear and vitally important for landing, apply the braking force.

However, all three hydraulic systems had to travel through the tail region to actuate the rudder and elevator surfaces within the empennage (tail structure).

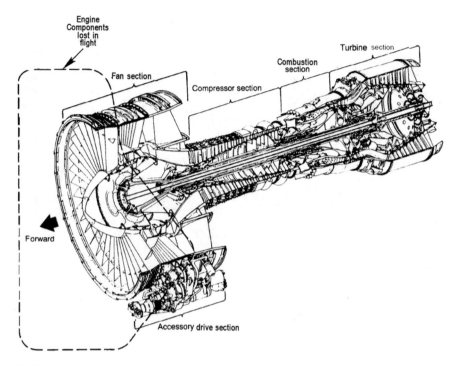

Fig. 9.4 Turbofan engine components

The explosive failure of the #2 engine that generated the difficulty in controlling the aircraft was caused by a burst fan disc. The fan disc is the large inboard, solid rotating assembly that holds the large fan blades in a turbofan engine. Figure 9.4 shows the general arrangement of the main sections of a large turbofan engine, like that found in the #2 slot for the DC-10. The #2 Engine position uses a much longer inlet duct than represented by Fig. 9.4, this is due to the engine's placement close to the aft end of the tail. The fan assembly is the rotating assembly at the front of the engine assembly, it provides the bulk of the thrust from the engine.

Figure 9.5 shows the tail region of the McDonnell Douglas-11, the next larger variant of the DC-10. The name change occurred due to Douglas being acquired by McDonnell. This model kept the same engine arrangement with the #2 engine mounted at the aft end of the duct that ran through the lower section of the tail rudder. The fan is the first rotating equipment to be encountered in the duct, mounted at the front of the engine. In a turbofan engine, it provides the predominant thrust from the engine, sucking in air at a rate of nearly 1300 pounds/second. Of that overall flow 1/6 enters the engine core (compressor, combustor and turbine) while the remaining 5/6 stays in the fan duct and exits the engine after passing through the fan. While engine manufacturers design to contain the energy in a lost fan blade, they did not (and probably could not) provide sufficient protection for the complete failure of the disc that housed all of the fan blades. Figure 9.6 shows the position of hydraulic lines in the tail area of the DC-10.

Fig. 9.5 MD-11 #2 engine location in tail region

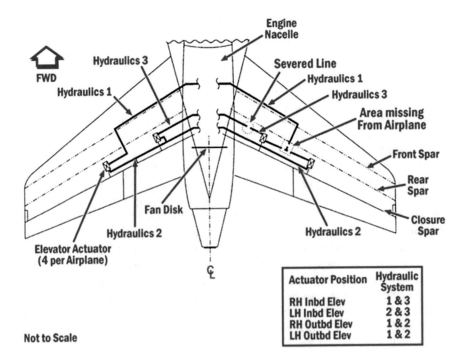

Fig. 9.6 Details of hydraulic lines in the DC-10 tail region

The immediate impact of the failed fan disc produced a severing of the #1 and #3 hydraulic system. In addition, the energetic fan disc release forced hydraulic components to be lost. These added components allowed the #2 hydraulic system to bleed its fluid too. The ultimate result was a complete loss of hydraulic fluid and the inability to move those flight control surfaces serviced by the hydraulic systems. Unfortunately, due to the size and airplane velocity virtually every important control function needed hydraulic pressure to actuate them.

The DC-10 was equipped with an air-driven generator (ADG) that could provide emergency electricity that could also power an emergency hydraulic pump. The crew deployed the ADG but noted no improvement in their ability to control the aircraft. The loss of all three hydraulic systems had drained all of the hydraulic fluid from the aircraft, leaving this backup hydraulic pump unable to return movement of the flight control surfaces to the cockpit.

A mere 4 min into the accident, the crew had declared an emergency and were told that Sioux City Iowa was the closest airport to their position. The immediate problem became how to turn the aircraft to the Sioux City airport. Without hydraulic pressure, there was no mechanism for moving the rudder, elevators, and ailerons, the control surfaces that normally provided direction, elevation, and roll control.

The pilot, copilot, and a United Airlines DC-10 training pilot (who happened to be riding as a passenger on the flight) used asymmetric thrust maneuvering to bring the DC-10 into a series of right-hand turns to align the airplane with the Sioux City Airport. This process constantly adjusted the power levels of the two wing-mounted engines to turn the aircraft. While not usually used, this process achieved different thrust levels from the engines that were mounted off the aircraft centerline. The pilots found that keeping the #1 engine (left) at 100% and the #3 engine (right) at 73% throttle would maintain level flight. Clearly some position of the rudder and/or other control surfaces that was now unmovable because of the damage from the engine failure and loss of hydraulic fluid generated a tendency for the aircraft to turn right. After the loss of the #2 engine, the plane would make only one left turn, and a series of right runs to align with the Sioux City Airport. The United flight manual contains zero instructions on how to operate the aircraft without hydraulic fluid. One DC-10 pilot noted after the crash, the only addendum to add to the flight manual for this type of accident is "Good Luck" [6]. It is considered an amazing feat of flying skill that the cockpit team was able to handle the aircraft as they did.

Sioux City airport contains three runways of varying lengths. Runways are numbered based on their compass heading so depending on which direction you approach them, you will see a different number, 180° (18) away from the opposite end. Runway 13/31 ran northwest/southeast, was an active runway 8999 ft. long, Runway 17/35 ran almost due north/south was also an active runway of 6599 ft. UAL flight 232 had lined up with runway 04/22 which ran northeast/southwest and was also 6599 ft. long. When the cockpit crew was notified that the runway was intact, but not active, they elected to continue their descent to that runway. The obvious difficulty they had in controlling the aircraft and their grave concern with the turns required to line up with a different runway were the major factors in that decision. Unfortunately, this inactive runway had become the staging area for emergency equipment for the anticipated difficulties expected upon the touchdown of UAL 232. With UAL 232 quickly approaching, the Sioux City control tower alerted the emergency equipment of the need to move off runway 22.

The Sioux City airport also housed a unit of A-7 Crusaders from the Iowa Air National Guard (ANG). The ANG base had an agreement with the local fire department to augment their units in the event of an emergency. This support could also be requested by the FAA from the control tower and was exercised for the United 232 crisis. A total of 5 airport fire trucks and 4 from the surrounding community were on-hand and ready at the airport before the touchdown of UAL 232.

While the flight crew were able to manually deploy the landing gear, the loss of all the hydraulic systems dealt the cockpit team a string of difficulties during their hoped for landing, and then stopping. Hydraulic pressure is used to deploy the slats and flaps that provide the aircraft with sufficient lift at the lower speeds desired for landing. Without the hydraulic systems, all of the subtle aircraft maneuvering that occurs just prior to landing would be unavailable. Finally, brakes for aircraft of this vintage are applied through hydraulic pressure acting upon the alternating rows of stators and rotors. The loss of hydraulic pressure would have left the aircraft without brakes, and the engine-mounted thrust reversers as the only viable method to slow the aircraft upon touching down. It is doubtful that thrust reversers could have stopped the plane before reaching the end of the runway, but the Sioux City airport air controllers were comforted by the level fields of crops that existed beyond the end of runway 22.

UAL flight 232 did not get to deploy their thrust reversers or plough into the fertile soil at the end of runway 22. Instead, the instability that had made for constant thrust adjustment in the cockpit reappeared just before touchdown. The pilots knew that they were traveling about 50% faster than a normal approach (~190 knots vs. typical approach speed of 130 knots) and would have no wheel-mounted braking system available. The aircraft sink rate was also about 10% higher than desired and the nose was beginning to pitch down. The pilots tried to increase the thrust to reduce the sink rate, and this action would also pitch the nose upward. Just before touchdown, as they tried to reduce overall airplane thrust, the imbalance in timing allowed the right wingtip to dip. This wingtip was the first point of the aircraft to touch the ground. It was followed by the #3 engine (right) and nacelle followed by the right-hand main landing gear and wheels. Unfortunately for the passengers and crew, the engine striking the ground initiated a fire in the engine area that spread to other parts of the aircraft.

The DC-10 began to cartwheel and break up after the large increase in drag started on the right-hand side from the wheels, landing gear, engine, and wing tip striking the ground. The right wing, tail, and cockpit all broke away from the main fuselage and came to rest at different locations near the runway. The fire spread to the main fuselage and despite the quick reaction of emergency personnel, many passengers died from the flames and smoke inhalation. The remaining fatalities were from blunt force trauma as sections of the fuselage came to rest along the airport. In the end, 185 passengers and crew survived the fiery crash and 111 perished (1 flight attendant and 110 passengers). The entire cockpit crew survived as the front of the aircraft broke away from the main fuselage and remained intact.

Root Cause

The CF6-6 engine that powered the initial DC-10 platform (−10 model) arose from a General Electric Aircraft Engine design the TF39 that powered the Air Force's C-5 cargo plane. The high-bypass turbofan engine used a fan of about 84 in. in diameter at the front of the engine to generate the large airflow required to achieve the 40,000 lbf of thrust that each engine could generate. The 8:1 bypass ratio CF6-6 engine became certified for commercial transport aircraft by the FAA in September 1970.

The FAA has long recognized the possibility of a fan blade being liberated from the fan disc assembly due to a bird strike or any other similar impact. The mass of a single blade rotating at nearly 4000 rpm could pierce the cabin or cause other severe damage to the aircraft and its occupants. Therefore, a fan shroud is built around the fan case to absorb this impact while protecting the aircraft and its passengers. However, absorbing the rotating energy contained in the central disc that retained all of the fan blades was not feasible, nor was it considered a credible accident scenario. The disc had been designed, gone through manufacturing, quality control and numerous on-plane inspections. These safeguards were thought sufficient to eliminate the possibility of the fan disc from fragmenting during flight.

The fan disc was made of a titanium alloy, a favorite among aerospace engineers because of its very high strength-to-weight ratio and superior corrosion resistance. The commonly used Ti6V4Al alloy offered over a 4× increase in strength vs. the common aerospace aluminum alloy 6061, with only a 60% increase in density. The ratio of those two factors (tensile strength divided by density), sometimes called specific strength or strength-to-weight ratio, was 2:1 in favor of the titanium alloy.

While titanium is a wonderful metal, with great applicability in the aerospace industry, it tends to be more brittle than other aerospace metals. This brittle nature has led to significant efforts to insure its purity during manufacture, in order to eliminate any inclusions that could initiate cracks. Any hard impurity within the titanium alloy could be a crack initiation site, plus the low elongation (brittleness) of the titanium alloy makes it more susceptible to cracking than other alloys. This inherent brittleness caused manufacturers to employ several steps to insure material purity and eliminate the potential for inclusions to act as crack initiation sites. GEAE initially used a double vacuum melt process for the titanium discs used, at the engine front face, to hold the fan blades. The vacuum melt process could reduce any hard inclusions that could occur due to the titanium's affinity to react with oxygen and/or nitrogen while in its molten state. GEAE formed the failed rotor disk in 1971, but instituted a triple melt process starting in 1972 to further reduce the probability of inclusions within the forged disc (Fig. 9.7).

The disc operated for 18 years (1971–1989) under thousands of cycles (16,999 takeoff/landing and presumably startup/shutdown cycles) before failure. However, GEAE had already instituted a new manufacturing method to eliminate the inclusion that initiated the crack that ultimately produced the accident. The National Transportation Safety Board (NTSB) cited human error along with this manufacturing process as a contributor to the accident. A total of 6 Fluorescent dye Penetrant

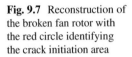
Fig. 9.7 Reconstruction of the broken fan rotor with the red circle identifying the crack initiation area

Inspections (FPI) had been performed on the disc while it was with United Airlines. The last inspection occurred 1 year before the accident (February 1988), 760 cycles before the accident. A post-accident analysis by GEAE estimated that the crack was at least 0.5 in. long at the time of that February 1988 inspection, easily large enough to be identified by a thorough FPI. It is perhaps ironic that the manufacturing defect that instigated the crack, that caused the crash of UAL 232, was eliminated 17 years before the accident. Only two potentially faulty rotors made by the previous double melt process were found and replaced [7].

McDonnel Douglas also implemented changes as a result of UAL 232. They integrated individual hydraulic shutoff valves and sensors to identify low hydraulic pressure in all three systems and lock pressure within the lines to prevent a reoccurrence of this type accident. While this system could not maintain the same level of pressure as normal, it would still allow pilots some minimal control, a much better situation than the one that the flight crew experienced on UAL 232.

References

1. Airline Reporter, Feb. 19, 2014, "A Historical Look at the DC-10 Before its Final Passenger Flight", Hull, K., http://www.airlinereporter.com/2014/02/historical-look-dc-10-final-passenger-flight/ retrieved 10/24/2017
2. Aircraft Accident Report, PB 219 370, NTSB, June 12,1972
3. FAA, Lessons Learned from Civil Aviation Accidents, Turk Hava Flight TK981, DC-10, http://lessonslearned.faa.gov/ll_main.cfm?TabID=3&LLID=18&LLTypeID=0, retrieved 10/25/17
4. FAA Lessons Learned from Civil Aviation Accidents, American Airlines DC-10, http://lessonslearned.faa.gov/ll_main.cfm?TabID=1&LLID=14&LLTypeID=0 retrieved 10/29/17
5. Flight 232: a story of disaster and survival, Gonzales, L., 2014, W. W. Norton and Co., ISBN 978-0-393-24002-3
6. No Left Turns, NASA System Failure Studies, Volume 2,Issue 6, July 2008.
7. Official Accident Report, NTSB/AAR-90/06, November 1, 1990

Chapter 10
In Flight Thrust Reverse Actuation

Lauda Air started as a charter service by Formula 1 driver Nikki Lauda. By 1991 it had expanded into scheduled revenue service, including routes to Australia with regular stops in Bangkok. While Lauda air started with 2 Fokker turboprops, it expanded into jets (Boeing 737 and 767) as turboprops could not provide the endurance needed for the Europe to Australia flight length.

The Boeing 767 is a wide-body (twin aisle) airliner that first flew on September 26, 1981, and then received FAA certification on July 30, 1982. Through 1990, there were no loss of hull accidents nor fatalities for the 767. Two 767s were destroyed shortly after the Iraqi invasion of Kuwait. Iraq pirated the two 767s from the Kuwait City airport to Mosul, Iraq. They were then destroyed shortly after the Allied bombing campaign began, which was part of the initial coalition attempt to force Iraq out of Kuwait. The Lauda air accident became the first fatal crash involving a 767.

Like most long-haul commercial transport aircraft built in the past three decades, the design relies on two (2) high-bypass turbofan engines. This model (Boeing 767) relies on engines delivering at least 48,000 pounds force of thrust (each) to power the airframe. Later 767 models which increased the fuselage length for greater capacity required slightly higher thrust levels. This twin-engine design allows airlines to reach lower maintenance costs vs. the early jet engine configurations that used three or four engines. Engine makers have achieved high thrust loaded engines which allow airplane designs to proceed with only 2 engines. Also similar to most Airbus and Boeing designs, the airframe has been qualified with engines from two different manufacturers to allow the airlines to compare offerings from different engine suppliers, generally insuring a competitive option. The 767 like most modern airliners has two engines slung under the main wings via a pylon, with the engine core just forward of the wing leading edge.

The Boeing 767 has been FAA and EASA qualified with both P&W JT9D and GE CF6 engines, in the 48,000–56,000 pound thrust class [1]. Boeing has produced over 1000 of the twin-aisle airframes in conventional airline configuration, as an all-freight variant and it is presently being developed as an aerial refueling platform for the US Air Force (KC-46A Pegasus). No new airline models are being built as it

© Springer Nature Switzerland AG 2020
T. Filburn, *Commercial Aviation in the Jet Era and the Systems that Make it Possible*, https://doi.org/10.1007/978-3-030-20111-1_10

has been replaced by the 787 Dreamliner in the Boeing twin aisle inventory for airline use. In revenue passenger service the 767 can carry between 215 and 290 passengers depending on the mix of seat class.

During the piston/propeller era, propeller actuation systems could change the propeller pitch (angle) on the hub depending on the flight condition (takeoff, cruise, etc.). This pitch change could even rotate the propellers sufficiently to produce reverse thrust, thus helping to slow the aircraft upon landing. With the advent of jet engines, transport aircraft have reached much higher weights, the Boeing 747 is about six times heavier than the Lockheed Constellation (one of the last large propeller driven transports) [2]. This greater weight coupled with the higher landing speed produces a much greater energy load on the braking system. While carbon/carbon brakes can absorb a significantly higher energy loading per volume and mass compared to the prior generation of steel brakes, they routinely need added help from jet engine thrust reversing. Thrust reversing upon landing is especially useful when the runway conditions are poor (ice, snow, or heavy rain) and hard braking may result in brakes locking up.

It is interesting that most civilian jet transport aircraft have opted for the installation of thrust reversers on their nacelles. It is also interesting when you consider they are in use for an extremely small period of time (10–20 s) out of 1–12 h flight time. They add weight, cost, especially the more complicated cascade system shown in Fig. 3.9c. The added flow surfaces and required steps and gaps in the nacelle also increase fuel burn, increasing TSFC by 0.5–1.0% [3]. Early in the adoption of TR systems, some planes could use them to back away from the gate area. The noise, limited visibility, and threat to ground crew make this a rarely used option, primarily available for smaller commuter aircraft operating at minor airports with less traffic and less ground equipment available.

The USAF C-17 can deploy its thrust reversers on the ground to back up, but the military airfields it is designed to service may not maintain the same suite of ground carts available at most commercial airports. The C-17 can also deploy its thrust reversers during flight to achieve an exceptionally short and steep landing path. Again, the exigencies of operating in areas with hostile forces make this feature not necessary for commercial transport.

Despite the negative factors listed above, airlines continue to ask for thrust reversers on their new plane acquisitions. While they may reduce brake wear, a stronger factor for their installation and operation relates to fouled runway operation. Airlines would generally be required to limit takeoff and landing weights on slippery runways, but the added braking force of the TRs allow them to maintain a more regular dispatch schedule plus typical landing load levels.

Most civilian jet transport aircraft can employ thrust reversing after landing. The cascade system used on the 767 with PW 4000 engines is common within the transport industry. Figure 10.1 shows a detailed cascade thrust-reversing system used on the Boeing 767 PW4060 engines, which operated with a 5:1 bypass ratio at cruise (5× more mass flow through the fan duct vs. the flow through the engine core). This figure shows both the reverser deployed (upper half) position and stowed (lower half) for normal flight.

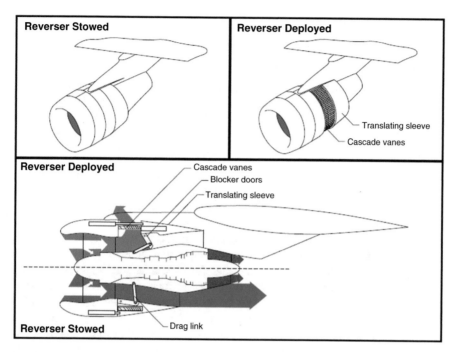

Fig. 10.1   Cascade-type thrust reverser deployed (upper) and stowed (lower) in high-bypass turbo-fan engine

Thrust reversers are not designed for in-flight deployment in civil transport. The Federal Aviation Regulation (FAR) states.

Each engine-reversing system intended for ground operation only must be designed so that in flight the engine will produce no more than idle thrust. In addition, it must be shown by analysis or test that

1. The reverser can be restored to the forward thrust position, or
2. The airplane is capable of continued safe flight and landing under any possible position of the thrust reverser.

While their actuation during flight was considered serious, it was also considered to be one that could be overcome by the prompt action of the cockpit crew. This assessment came from a history that started with predominantly 4-engine craft (e.g., Boeing 707) and then continued with lower speed testing on the 2-engine 767. As shown due to the high-speed deployment of 1 thrust reverser on Lauda Air NG 004, this assessment was wrong. In fact, post-crash simulation indicated that well-trained pilots on the 767 type had 4–6 s to respond to the inadvertent TR deployment or recovery of controlled flight could not be achieved. The genesis of this faulty belief can be traced to several inaccurate assumptions, and extrapolation of data that did not align well with the 767 installation.

These faulty assumptions included:

1. A Boeing 747 had experienced in-flight reverse thrust, the failure in extrapolating this data to the 767 was twofold,

   (a) The 747 was a 4-engine craft; therefore, reverse thrust on 1 engine produced significantly less yawing moment on the aircraft vs. the 2 engine 767.
   (b) The 747 pylons were longer vs. the 767, this meant that any reverse thrust plume was further from the wing and had less chance to interfere with flow over the wing and has less impact on wing lift.

2. Boeing had wind tunnel data that was used to create a model of the thrust reverser in-flight actuation. This data was only complete up to 200 knots, less than ½ the velocity of Lauda Air NG 004. The higher speeds of the aircraft created much higher impact on wing lift.

3. Most in-flight testing of TR actuation had occurred with the engine set for idle thrust, the Lauda Air flight actuation occurred during climbout when the engines were set for a high thrust level and therefore TR actuation would produce an immediate and significant reverse thrust level. These higher thrust levels produced two significantly larger impacts on the plane operation vs. the assumed flight characteristics. The first impact placed a much larger yaw (twisting) moment on the aircraft with the right engine still generating significant forward thrust and the left engine providing significant reverse thrust. The second impact was the large loss of lift on the left wing due to the large flow disturbance from the reverse flow plume over the wing.

Lauda Air flight NG 004 was on its regularly scheduled service, Hong Kong to Vienna, Austria, with a planned stop in Bangkok, Thailand. The flight left Bangkok airport at 16:02 local time on May 26, 1991. The flight operated on a Boeing 767-300 ER (extended range), favored by many airlines for long distance routes, like the Hong Kong to Vienna service. The weather in Bangkok was good, with visibility on the ground extending for 6 miles, light winds of 6 knots, and cloud cover of about 1/3 of the sky in the airport vicinity. Not surprisingly for Bangkok in May, the temperature was near 80 F, and the relative humidity was just over 80 percent.

The pilot of Flight NG 004, Thomas Welch had approximately 11,750 of flight time in turbojet aircraft. His copilot, Josef Tuner had about 6500 h. The flight carried 213 passengers along with 8 additional crew members as it readied for takeoff from Bangkok airport for its Vienna destination on that fateful Sunday afternoon. The flight suffered a breakup in air about 15 min after takeoff and crashed into mountainous jungle terrain. Emergency and accident recovery personnel found no survivors. In addition, the digital flight data recorder (DFDR aka "the black box") was sent to the USA for data recovery, but either the in-flight breakup or ground fire made the medium unreadable and no data was retrieved. The predominant data source was therefore the cockpit voice recorder (CVR) and the electronic engine control (EEC) along with data from the engine manufacturer (Pratt & Whitney) and engine controller manufacturer (Hamilton Standard).

All indications from the post-crash investigation point to a normal routine preflight, ground travel, and takeoff. However, the cockpit voice recorder identified a crew alert associated with a thrust reverser isolation valve. The crew consulted their Quick Reference Handbook (QRH) to determine their response, with no actions taken nor any required by the QRH [4].

Takeoff time was derived from the noise of the engines spooling up, that sound can be clearly identified on the CVR. The flight appeared to be normal until 5 min and 45 s after takeoff. At that time, the cockpit crew can be heard discussing the illumination of the REV ISLN annunciator (reverser isolation). This annunciator could represent either a REV ISLN yellow light in the center pedestal, or an L REV ISLN VAL (left reverser isolation valve) yellow caution light from the engine indication and crew-alerting system (EICAS) or both lights could have illuminated. The crew discussed these alerts for the next 4 ½ min but took no action. 10 min and 20 s into the flight the copilot advised of the need for rudder trim to the left, which was acknowledged by the pilot. 15 min and 1 s into the flight, the copilot states "ah left reverser's deployed." Sounds akin to the airframe shuddering can be heard on the CVR. 29 s later the CVR recording ended with the sounds thought to be breakup of the airframe. Information from the electronic engine control coupled with the CVR indicate that the left engine thrust reverser deployed while the airplane was at 24,700 ft., Mach 0.78 climbing to its cruising altitude of 31,000 ft. It is clear based on the copilot's pronouncement that the flight crew recognized this event, but that the airplane departed controlled flight, accelerated past its maximum operating velocity and experienced an in-flight structural failure [5].

The cascade-type thrust reverser used with the PW 4000 engines on this Boeing 767 relies on a moving outer housing (translating sleeve) to expose the cascade louvers that redirect the fan flow into the reverse direction, thus providing reverse thrust and slowing the aircraft. In most configurations, the translating sleeve is linked to a set of multiple (usually 6 or more) hidden (blocker) doors in the fan duct. The aft motion of the translating sleeve moves these doors from their stowed position and provides a large flow blockage to the aft movement of the fan air. The air is therefore redirected through the cascades, providing a force to slow the aircraft. Figure 10.2 shows these sections of the cascade thrust reverser. What is not shown in Fig. 10.2 is the prime mover, the fundamental mechanism that drives the translating

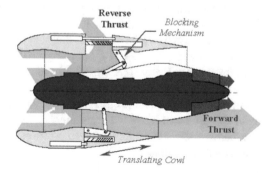

Fig. 10.2 Cascade thrust reverser showing translating cowl (sleeve), blocker doors, and exposed (top) and stowed (bottom) configurations

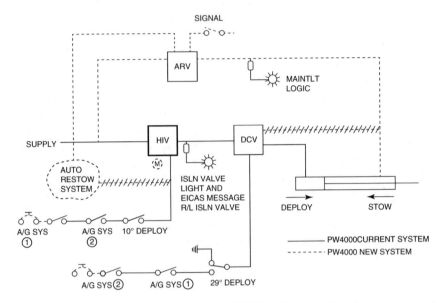

Fig. 10.3 Control signals and hydraulic lines for B767 Thrust Reverser Actuation

sleeve rearward, and through its linkages moves the blocker doors into their aft location for obstructing normal rearward fan flow. In the 767 design and many of that same vintage, hydraulic pressure provided the power to move the translating sleeve and blocker doors. In the 787 and many recent designs, the effort to more electric aircraft has shifted this prime mover to an electric motor.

The key to the inadvertent movement of the left engine thrust reverser in the Lauda flight, lies in the design of the hydraulic actuation system. This system, along with its electric switches and solenoid valves, is shown in Fig. 10.3. Ultimately a high-pressure hydraulic line would be opened to push the traveling cowl aft and exposing the cascades to fan flow. If the engine was at a high rpm, regardless of flight regime, it would provide significant reverse thrust to the aircraft.

The intentional method to command the thrust reversing system (TRS) to deploy requires the pilot to lift the reverse-thrust lever in the control stand of the flight deck, by more than 10 °. This action will close the switch marked 10° deploy (see yellow highlighted switch) and will open the solenoid (electromagnetic coil) operated hydraulic isolation valve (HIV) allowing pressurized hydraulic fluid to reach the directional control valve (DCV). Additional movement of the thrust-reversing lever in the cockpit (at least 29°) closes the directional control valve switch and opens the directional control valve itself. When both valves have been opened, high-pressure hydraulic fluid will move the translating sleeve to the deployed position. When the translating sleeves leave their stowed position, the reverser's in-transit light (REV, amber) will illuminate [6].

What Happened on Lauda Air NG004

The 767 PW4000 thrust reversers like almost all non-military TRs are designed to be engaged during ground operation only. It is extremely unlikely, and the CVR seems to agree that the flight crew did not intentionally engage the left engine TR. As mentioned above, two distinct actions must be completed that power two hydraulic valves that are situated in series. The 767 PW4000 TR system has several layers of protection, designed to prevent the inadvertent actuation of the TR system in-flight. Unfortunately, the postaccident investigation discovered that if certain abnormal conditions existed with the auto-restow circuitry during flight, this abnormal condition could circumvent the protection designed to prevent in-flight deployment.

Investigators including the FAA and NTSB from the USA examined whether an electrical system failure could result in an uncommanded deployment of the TR. This investigation uncovered the possibility that an electrical short could command the DCV to move into the deploy position, if it received an auto-restow command. This counterintuitive result would obviously surprise any flight deck crew.

In order to achieve any movement of the TR system, the Hydraulic Isolation Valve must first be opened to permit hydraulic fluid pressure to reach and operate the rest of the system. The HIV can be opened by a circuit within the air/ground sensing system (e.g., the aircraft is on the ground) or through the auto-restow circuit. This type of electrical failure is not unheard of, other B767 experienced the REV ISLN indication due to wiring anomalies, meaning that the auto-restow circuit annunciated during flight.

Because the FDR was unreadable, the majority of the information presented is based on the CVR, the electronic engine control system data, and recovery of the aircraft pieces on the ground. It is also based on a new engineering simulation that the Boeing Commercial Airplane Group developed after the accident, to simulate the in-flight deployment of a TR while at high power and near cruise velocity.

When the TR deployed at high speed (M = 0.78) and high thrust on both engines, a large yaw load was immediately put on the airplane, along with a significant loss of lift for the port (left) wing where the TR deployed. This forced the aircraft into a roll, which required immediate cockpit response. The modified Boeing simulation that included a more pronounced effect from TR deployment found a recovery time of only 4 s, which could be 6 s if the pilot immediately commanded the right (starboard, non-TR deployed engine) to idle. Otherwise the combined aerodynamic loss of lift and power yaw produced a left roll rate of 28° per second, generating a left bank in excess of 90° within 5 s.

Based on the data from this new Boeing computer model of the airplane flight characteristics and the recovered information, the aircraft appeared to suffer an in-flight breakup due to maneuvering overload and excessive speed. It seems that the rudder and most of the left elevator separated initially. The right elevator then separated, leaving the aircraft without most of its tail, in a downward rapid descent. The overspeed condition on the wings ripped them from the fuselage, and an in-flight fire occurred as fuel from the wing tanks ignited. All of this destruction happened over a period of seconds (probably <10 s). No shrapnel or explosive residue was found, indicating that this failure was not the result of a hostile act (explosive or missile attack), but did result from the uncommanded deployment of the port (no. 1) engine thrust reverser.

Postaccident Changes

By July 3, 1991 (5 weeks after the accident), the NTSB made several specific recommendations in an official letter to the FAA [7].

1. This letter requested the FAA to review the certification of the Boeing 767 with the P&W 4000 engine to evaluate problems that can allow the TR to deploy inadvertently.
2. It suggested amending the 767 Flight Operations Manual to include a caution on the severity and danger that can follow illumination of the "Reverser Isolation Caution Light."
3. It recommended Boeing and the FAA establish operational procedural changes to be followed if the Reverse Isolation Caution Light (REV ISLN) illuminates.
4. It recommended a thorough certification basis review for similar (presumably 2 engine transport) aircraft with electrically or electro-hydraulic actuated thrust reverser systems for appropriate safeguards to prevent in-flight deployment.

The NTSB listed items 1–3 as Class I, Urgent Action and item 4 as a class II, Priority Action.

The FAA and Boeing responded to the NTSB letter with both immediate changes and a long-term study. Through Air Worthiness Directive T91-18-31 (effective 8/23/91), all thrust reversers were deactivated on 767s powered by PW4000 engines.

Boeing collaborated on a redesign of the TR system for all the engine systems on its 767 aircraft, as well as examining the TR systems for all of its commercial aircraft. It worked on redesign efforts to insure that none of these aircraft had a single point failure that could allow the TR system to be deployed in-flight.

The FAA (and EASA) restructured their certification process to require demonstration that the TR system will not be deployed during any phase of flight. All commercial transport aircraft now have at least 2 interlocks to prevent the inadvertent deployment of TRs during flight.

The FAA also began to investigate the requirements for the survivability of flight data recorders. The loss of the Lauda FDR due to fire became another lost FDR, with critical flight information. The time/temperature requirements that these FDRs had been designed to survive were clearly inadequate as evidenced by the lack of survivability for this FDR and additional ones from previous accidents.

References

1. http://www.boeing.com/history/products/767.page, retrieved 5/13/2018
2. Quest for Performance, the evolution of modern aircraft, L. Loftin, jr, NASA 1985
3. Why Do Airlines Want and Use Thrust Reversers, NASA TM 109158, Yetter, Y. A., 1995
4. Lauda Air B767 Accident Report, Ministry of Transport and Communications Thailand, approved July 21, 1993
5. National Transportation Safety Board Memo, A-91-45 through −48, July 3, 1991
6. https://digital.library.unt.edu/ark:/67531/metadc712270/m1/1/ accessed 8/3/18

Chapter 11
Cabin Pressurization Accident

Boeing's model 747 marked a new chapter in commercial aviation history. While its design can be traced to Boeing losing the US Air Force's request for a large, long-range transport (won by Lockheed, now Lockheed Martin), the C-5, late in 1965. However, that loss coupled with the growth in airline passenger traffic spurred Boeing into developing a commercial aircraft with a load capacity well beyond anything then in the inventory of US and European carriers.

Boeing would take advantage of the new high-bypass high-thrust turbofan engine technology that had been developed for the C-5. GE's engine type for the C-5, the TF39 could develop over 40,000 lbf of thrust, vs. the 21,000 lbf thrust of P&W's JT8D, which operated with a 1.74:1 bypass ratio. The large thrust capacity of GE's TF39 and a short time later P&W's JT9D allowed Boeing and other airframe design companies to conceive of a new class of large passenger airplanes. The large thrust capacity, coupled with the reduced TSFC of these new engines, meant that these large airplanes could compete with, or operate with, reduced seat-mile costs. The high thrust meant that a 4-engine configuration could be designed, which eliminated the extra drag and higher maintenance from having 6 or 8 (e.g., B-52) engines to provide thrust.

Boeing was the first OEM to offer a wide-body high load (seating) capacity air-frame, with its 747-100 model, which had its flight in 1969. Boeing has produced later models (-200, -300, -400, and -8 an all freighter configuration) that can accommodate nearly 500 passengers. Its inaugural -100 model could hold 420 passengers in a two-class configuration. The -100 model could lift-off with a max takeoff weight (MTOW) in excess of 700,000 lbm.

United Airlines is a US airline, with one of the longest continuous operating histories. United Airlines actually operated as a part of the Boeing aircraft company in the 1920s. In fact, the United Aircraft and Transport Co. was formed in 1929. This company included Pratt &Whitney Aircraft and Boeing. In 1931 it combined with several small airlines to create the United Air Lines, a vertical monopoly that built the engines, the airplanes (even the propellers), and then flew them on revenue routes. By 1934, the US government forced the company to divide into three

© Springer Nature Switzerland AG 2020
T. Filburn, *Commercial Aviation in the Jet Era and the Systems that Make it Possible*, https://doi.org/10.1007/978-3-030-20111-1_11

separate companies, United Aircraft Co., Boeing Airplane Co., and United Air Lines. United Aircraft Co. would eventually become United Technologies Corp., the present conglomerate that operates P&W aircraft engines (ranking 51st on the Fortune corporate list by sales). Boeing has grown into the world's largest aerospace corporation, putting it 27th (in 2017) on Fortune's list of public companies. Finally, United Airlines has grown into the 81st largest corporation according to the 2017 Fortune list. Today the airline operates more than 4500 flights per day, to 338 airports. It flies more than 1200 aircraft, including over 500 regional aircraft (0.95% jets) and 744 single-aisle and wide-body aircraft from Boeing and Airbus. Although not the launch customer (Pan Am 1969) United started flying the 747 in 1970 and continued with this type until 2017. It finally retired the 4-engine wide-body as newer 2-engine aircraft (e.g., 787) which relied on engines with higher thrust and lower TSFC which promise lower cost (seat passenger miles) especially on long distance transcontinental flights.

Like most new airplane types, the 747 had some initial teething pains when it was first introduced to service in 1969. However, the first fatal incident with a 747 would occur over 15 years after its maiden voyage and would be the result of a terrorist attack. Air India Flight 182 blew up on June 23, 1985, over Irish airspace killing 328 passengers and crew during a Montreal to London flight. A militant Sikh group was believed responsible for the bombing that brought down this aircraft.

The first US airline to suffer fatalities and a hull loss flying the 747 was Pan Am flight 103 which was bombed during a London to New York flight. This was also the result of a terror bombing and occurred on December 21, 1988. All 259 passengers and crew were killed along with 11 UK citizens caught under the debris field on the ground near Lockerbie Scotland. In 2000 two Libyans were convicted of building and placing the bomb on this Pan Am flight.

The first fatal crash of a 747 not associated with a terrorist attack occurred on Feb. 19, 1989. A Flying Tiger 747, flying a cargo mission from Singapore, with 3 crewmembers and 1 passenger crashed on approach to Subang Airport in Kuala Lumpur Malaysia. The aircraft hit a hillside several miles short of the runway after numerous communication and procedural errors on the part of the flight crew during approach.

On Friday, February 24, 1989, United Airlines Flight 811 began its scheduled departure from Los Angeles International Airport (LAX). This flight had planned stopovers in Honolulu Hawaii (HNL) and Auckland New Zealand (AKL) before it would reach its final destination of Sydney, Australia. The crew flying the initial LAX to HNL leg reported no anomalies with the flight or aircraft. However, a new crew would fly the Boeing 747-100 model from HNL to SYD. The aircraft departed Honolulu at 01:33 am local time (3 min late) for its expected 10 ½ h flight to Sydney. The aircraft left HNL with 3 flight crew, 15 cabin crew, and 337 passengers.

The second officer indicated that all cabin and cargo warning lights were out prior to the airplane leaving the gate. The 4 main engines along with the APU were all operating during the takeoff roll. The APU was shut down shortly after the main engines were set to the lower power down climb thrust from their higher power takeoff setting.

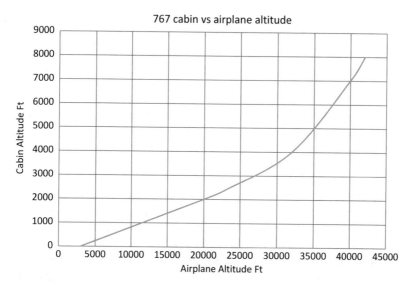

Fig. 11.1 767 Cabin altitude schedule

As Flight 811 climbed to 22,000 ft. altitude, the cabin ECS system would have been continually bleeding some outside air from the compressor section of the engine into the cabin and cockpit. This air is part of the FAA mandated 0.55 lbm of fresh air makeup per occupant [1]. The cabin pressurization system would have begun operating shortly after takeoff to limit the rate of pressure change within the cabin. This pressurization system can control the 747 outflow valve to maintain a minimum pressure within the aircraft, while also insuring the necessary fresh air makeup. For this vintage 747, the aircraft would try to maintain the pressure at an equivalent altitude of 6000–10,000 ft. altitude once the aircraft had climbed to this altitude. Figure 11.1 shows a cabin altitude chart for a similar Boeing model 767. In order to limit passenger discomfort, the pressurization system operates during most of the climb phase, limiting the time rate of pressure change, the minimum pressure inside the cabin, and the overall pressure difference (inside-to-outside pressure) [2].

While climbing away from the airport, the pilot noted thunderstorms in the area, requested a course deviation to avoid them, but elected to leave the passenger seat belt "on" sign illuminated. United 811 climbed at a rate of 2000–4000 ft. per minute after leaving Honolulu, it reached 22,000 ft. altitude about 15 min after takeoff. At this time, with the aircraft traveling at 300 knots, the flight crew heard a thump, which shook the airplane. This sound was immediately followed by a "tremendous explosion."

The pilot and copilot immediately donned oxygen masks but found no oxygen available. The airplane cabin altitude horn sounded (low cabin pressure) and the flight crew believed that the cabin oxygen masks deployed automatically.

The captain began an emergency descent, turning left to avoid the local thunderstorm and completed a 180° turn to return to HNL. The first officer informed the

HNL tower that flight 811 was in an emergency descent and appeared to have lost power in the No. 3 engine (inboard engine on right wing). Shortly after this, the No. 3 engine was shut down because of heavy vibration, no N1 fan/ low compressor rotation indication, low exhaust gas temperature (EGT), and low engine pressure ratio (EPR). As noted in Chap. 7, there is no direct measurement of thrust generated by an engine. These alternate sensors (rotation speed, turbine exhaust temperature, and engine pressure ratio) act as surrogates and together provide insight into the engine performance, which now indicated this engine was not generating thrust.

The second officer left the cockpit to inspect the cabin, returning to inform the captain that a large portion of the forward right cabin fuselage was missing. The captain subsequently shut down the number 4 engine (outboard engine on right side), because of high EGT and no N1 compressor indication, in addition fire could be seen intermittently around the engine. After this the flight crew began to dump fuel as they exceeded their safe max landing weight due to their large fuel reserve for their planned long duration flight to Auckland.

HNL control cleared the flight for approach to runway 8 L. Runways are numbered according to the compass direction they face in 10° increments, so 8 L represented 80° from north which would be nearly a due east runway. The L designation represented the left hand of parallel runways, with runway 8R located about ¾ mile south of 8 L. Because of the compass heading, approaching the runway from the opposite direction, it would be labeled 26R. 18 (180°) higher. This runway numbering system allows pilots to quickly understand the impact of wind (tail, head, or crosswind) during the takeoff and landing process.

The final approach to runway 8 L was flown at 190–200 knots, using only the 2 left-hand engines, No. 1 and No. 2. Not surprisingly, the flight crew noted asymmetrical flap positions as they tried to extend them for greater lift during landing. The same debris that damaged both right-hand engines impacted the flaps and/or their actuators on the same side of the aircraft. The flight crew limited the trailing edge flaps to 10° and the right leading edge flaps did not extend for the landing. The suboptimal flaps position lead to a higher landing speed (typically ~160 knots). Again due to both starboard engines not operating, the captain did not use engine thrust reverse upon landing because of the large yawing moment that would have been produced. Instead he relied on heavy braking and brought the aircraft to a stop with nearly ¾ mile of runway still available. After stopping the aircraft, an emergency evacuation commenced on runway 8 L [3].

After the evacuation an examination of the aircraft revealed a large hole on the right side of the fuselage. This open area (~10 ft. × 15 ft) was in the forward lobe area (distinctive 747 hump) and extended down to the forward cargo door, which was missing. The lower door sill and side frames remained. Figure 11.2 shows an image of the damaged area after the aircraft's safe return to HNL.

An examination of the failed area disclosed no cracks or corrosion. The fracture surfaces appeared shiny indicative of fresh damage induced by an overstress condition. Debris from the door or fuselage had damaged portions of the right wing, the vertical stabilizer, and as previously noted engines No.3 and No. 4.

Fig. 11.2 Damaged
fuselage and missing cargo
door UA 811

The right wing damage included leading edge surfaces and the leading edge flaps. Numerous scrapes, dents, and gashes were found on the wing's leading edge, upper and lower surfaces. In addition to the dents, through punctures were noted on fairings that supported several flap surfaces. Finally, the pneumatic duct that deiced the leading flaps had been severed by a piece of metal. It is not surprising that the leading edge flaps did not deploy properly on this wing due to the amount of impact damage.

The right-hand stabilizer had several dents in its leading edge. The largest dent was 3 in. wide and nearly 1 in. deep, but no punctures were noted. The No. 3 engine nacelle had numerous FOD, including tears, scuff, and a large outwardly directed hole. All the fan blades on the No. 3 engine exhibited extensive FOD on their leading edges. The No. 4 engine nacelle had lesser scuff marks (vs. No. 3) and the engine fan blades had sustained both soft and hard object damage.

When the cargo door separated it took a section of the fuselage shell above it, along with a section of the main cabin floor below seats 8GH- 12GH. Those 10 seats (8G, 8H, 9G, 9H, 10G, 10H, 11G, 11H, 12G, and 12H) are presumed to account for most of the 9 fatalities who were ejected from the aircraft and lost at sea when the door, fuselage skin, and floor section left the aircraft. The oxygen supply and fill lines for the flight crew storage cylinder had been broken below the cabin floor, explaining why the oxygen masks for the cockpit were inoperative when donned after the accident. The UAL cost estimate to repair the aircraft and return it to revenue service (which was completed) was $14,000,000.

Failure Investigation

The postaccident investigation focused on the forward lower cargo door, which had departed the aircraft. The issue centered around the design, operation, and maintenance of these doors which were required to maintain cabin pressure and cabin safety. The forward and aft cargo doors are similar in appearance and operation on the 747. The door opening in the fuselage is approximately 110 in. wide, by 99 in. high, large enough to handle shipping pallets which now measure 64 in. high by 60 in. wide. Both doors have upper hinges and lower latches to allow the door to swing up and away, improving access to the cargo hold within. The cargo door switches and actuation are supplied by a ground handling electric bus, which can only be powered by an APU or external (ground) power. The electric generators on the engine cannot supply power to the ground handling bus. If the APU is started in-flight, the air/ground safety relay automatically disconnects the APU generator from the ground handling bus circuit.

The cargo doors are meant to handle the loads associated with the internal pressure of the aircraft against any expected flight altitude. An internal equivalent altitude of 6000 ft. and a typical flight altitude of 38,000 ft. would produce an inside/outside differential pressure of 8.78 lbf/in^2. While this seems like a small number, this pressure is uniformly applied across the entire 10,000 + in^2 of door area, producing an outward force greater than 45 tons. To resist this force, the door relied on piano hinges (a continuous hinge along the top) and 8 latches located along the bottom of the door. Even at the 22,000 ft. elevation, and assuming a slightly higher inside equivalent altitude of 8000 ft., the differential pressure would be 4.7 lbf/in^2, a still large force of 45,000 lbf on the door. Figure 11.3 shows a setup of the door assembly with the hinges, latches, and actuation mechanism.

Along with the top continuous hinge and 8 lower latches, the lower cargo doors also had 2 mid-span latches at the fore and aft location to keep the sides of the door

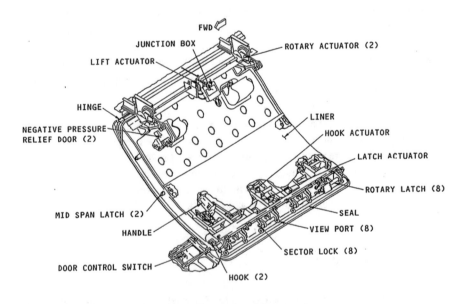

CARGO COMPARTMENT DOOR

Fig. 11.3 747 Lower cargo door system

Fig. 11.4 Cam lock
mechanism for cargo door
latches

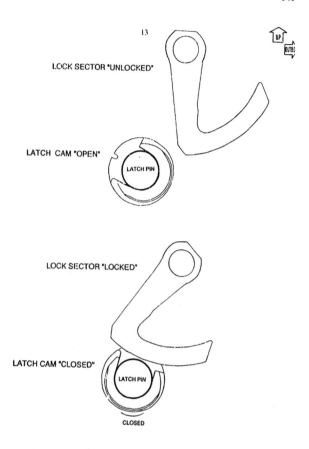

aligned with the fuselage. Four door stops limited the inboard travel of the door. The door mechanism had two pull-in hooks at the lower corners of the door.

Ground personnel can operate the door via an exterior switch, just forward of the door. Moving this switch to the "open" position will operate three electric actuators in succession. It will initially rotate the latch cams (see Fig. 11.4) to the open position (1), the switch will then move the pull-in hooks to the open position (2), and then finally the main actuator will move the door to the fully open position (3). Moving this switch to the "closed" position will reverse these operations. Ramp cargo handlers can operate the door manually, but the ramp personnel indicated that they had used the electric actuation for both the cargo door opening and closing iterations.

As the cargo door seemed to be the key evidence in the failure investigation, the NTSB encouraged the US Navy to identify the location and retrieve the door if possible. The Navy was able to identify the door and fuselage skin location during the summer of 1990. Recovery operations were completed in the fall, with the retrieval of the lower door half, along with the upper door half and a fuselage section. All of this debris was retrieved from the ocean floor, at a depth of approximately 14,200 ft.

Retrieving the door allowed investigators to rule out fatigue and corrosion as potential failure mechanisms. Examination of the door showed it to be fractured near the mid-span with no evidence of progressive fracturing on the door surface. All the cargo door lock sectors were in the locked position, but that the latch cams were nearly in the open position. Detailed examination of the latch pin surfaces found wear and damage.

As frequently found in these types of failures, warning signs had gone unheeded. Pan Am flight 125 from London to New York experienced an incident with the same forward lower cargo door, on its flight of March 10, 1987. The flight crew experienced pressurization problems as the airplane was climbing through about 20,000 ft. The crew began to descend, and the problem abated about 15,000 ft. The crew began to climb again, but again around 20,000 ft., the cabin pressure began falling (cabin equivalent altitude rising) rapidly. The flight returned to London. Examining the airplane on the ground, the forward cargo door was found ajar, about 1 ½ in. along the bottom where the latch cams had unlatched, but the master latch lock indicated closed. Even more troubling, the cockpit cargo door warning annunciator remained unlit. In this instance, the cargo door had been closed manually, but had been back-driven open by the manual socket drive system.

Postaccident Actions

Following the United accident, the FAA issued an Airworthiness Directive ADT 89-054, which required operators to follow specific procedures when operating the cargo doors on the Boeing 747. This directive also required rechecks on the doors and its locking mechanism. The directive also accelerated the planned placement of steel doublers on the latch lock sectors.

A key feature of the 747 cargo door was its non-plug design. That is the door simply opened directly outward, to facilitate cargo loading and unloading. This design is distinctly different from the traditional passenger doors which first open inward because they are sealed against the fuselage with the cabin pressure helping to keep them closed. These cargo doors had cabin pressure attempting to force them open, with only the latch mechanism keeping them pressure tight. The loss of a cargo door can produce serious aircraft damage (UA 811), the NTSB recommended the FAA verify the safe design and operation of all non-plug cargo door designs, beyond those installed on the 747.

By November 1990, Boeing had designed a new door latch switch for all the subject 747 cargo doors. They issued a Service Bulletin to operators for installation of this new safety device. The FAA allowed operators 18 months to implement these changes, but no new failures occurred during this installation window.

References

1. 14CFR 25.831 Airworthiness Standards: Transport Category Airplanes, Subpart D, Ventilation
2. Space, D., Johnson, R., Rankin, W., Nagda, N.,,"The Airplane Cabin Environment: Past, Present and Future" Air Quality and comfort in airline cabins, ASTM STP 1383, 2000
3. Aircraft Accident Report, UA 811, February 24,1989, NTSB AAR-92/02

Chapter 12
Landing Gear Accident

Many airlines initially signed up for the Mach 2 (1300 + mph, twice the speed of sound) speed of supersonic airliners in the 1960s and 70s which was >2× the speed of current airliners (~575 mph). Boeing was well along on design work for a commercial supersonic transport (SST) albeit with significant government funding for their first full-scale prototype. The Boeing model was designed to hold 300 passengers and travel at Mach 2.6, until concern about noise and pollution led the US government to abandon their financial support. The elimination of government funding forced Boeing to cancel the entire program [1].

The British and French had reached an agreement in 1962 to jointly develop an SST which they continued even after Boeing's cancellation. This agreement led to the production of the Concorde SST, which had its inaugural flight on March 2, 1969, exceeding Mach 1 by October of that year, and achieving its design flight speed of Mach 2 by November 1970.

The UK and French agreement would eventually produce 20 aircraft, with 6 being prototype aircraft for demonstrating the numerous subsystems that would differ vastly than those same subsystems found on their subsonic brethren. These differences emanated from both the much greater speed of the Concorde, its delta wing design, and operating altitude (60,000 ft). The 14 revenue aircraft were equally split between British Airways and Air France, entering revenue service in 1976.

While only a small fleet existed, the aircraft introduced technology into commercial Mach 2 flight normally reserved for military aircraft. The Concorde used a droop nose to allow pilots forward and downward visibility while taxiing, taking off, or landing. The large delta wing of the aircraft meant that it would land with a high angle of attack, without the droop nose system, pilots would have no visibility of the runway during approach. This same droop nose system was designed into the prototype supersonic US Bomber, the XB-70 Valkyrie during the 1960s.

The high-speed flight of the aircraft, and its continued time spent at these high speeds, meant that skin friction would heat the aircraft and cause it to stretch nearly 10 in. (out of 200 + ft. overall length) during its flight. Concorde would use heat resistant titanium and special white paint to limit the temperature rise, and retain strength during flight.

© Springer Nature Switzerland AG 2020
T. Filburn, *Commercial Aviation in the Jet Era and the Systems that Make it Possible*, https://doi.org/10.1007/978-3-030-20111-1_12

The US SR-71 spy plane which flew at Mach 3+ also was designed for large thermal growth of metal surfaces. Because the SR-71 flew so high, it used black paint to help radiate the skin friction back to the space and the surrounding atmosphere.

Concorde relied on 4 Rolls-Royce/SNECMA turbojet engines to generate 38,000 lbf of thrust (each) reach its high takeoff velocity as well as take the airliner through the sound barrier on its way to Mach 2. The engines were buried in nacelles directly below the delta wing. The inlet nozzles were carefully crafted to slow the incoming air down (even at supersonic speeds the inlet air never exceeded Mach 0.5), thereby preventing shock waves that could damage the engines or produce an in-flight shutdown. The engines could and did use augmented thrust (afterburners) in which raw fuel was dumped into the exhaust, thus providing a significant boost in thrust. This augmentation could be used at takeoff and to push the aircraft from Mach 1 up to its cruising speed of Mach 2. Most military fighter aircraft rely on the same carefully crafted inlet duct to slow the incoming air and these same fighter airplanes routinely use afterburners to achieve high levels of acceleration [2].

With such a fast aircraft, the Concorde flew even higher than conventional jet airliners. The extreme altitude allowed Concorde to fly through less dense air vs. its slower brethren. The Concorde was certified to fly up to 60,000 ft. and routinely flew at 55,000 ft., well above the 35,000–40,000 level for the rest of commercial aviation. The higher altitude represented lower pressure and therefore lower density air thereby reducing drag a key parameter in reaching their design mission lengths.

The compromises that had to be designed into the Concorde to allow it to fly at its Mach 2 cruise speed impacted other flight regimes beyond its high-altitude, high-speed cruise. Its droop nose allowed pilot visibility during takeoff and landing. Its delta wing, while imperative for high-speed travel, made it difficult to incorporate the same type of lift enhancement control surfaces (leading edge and trailing edge flaps) found on most other turbofan-powered airliners. Without these devices, the Concorde had to rely on a higher takeoff and landing speed vs. conventional subsonic airliners. The Concorde routinely rotated its nose off the ground at 200 mph while its main wheels would take off at 250 mph, well above that for a typical subsonic airliner 150–170 mph. The same lack of lift at low speed led to a high landing speed for Concorde, around 190 mph vs. a 747 landing speed of 145 mph. The Concorde operated with extremely high pressure tires 191 psia to minimize runway friction and to handle the overall takeoff weight. This air pressure was higher than most aircraft tires for similar weight aircraft which are in the 125–150 psia range; however the high takeoff and landing speed meant that the designers wanted to limit rolling resistance and would select fewer, but higher pressure tires to support the aircraft weight during its takeoff and landing roll. Finally the aircraft incorporated a smaller pair of wheels near the tail to protect the fuselage from striking the ground during takeoff because of the higher rotation angle taken by Concorde.

The sun rose to a partly cloudy dawn on July 25, 2000, at Paris's Charles De Gaulle (CDG) airport. The day would be pleasant with low humidity and temperatures in the mid-70s. Air France had chartered one of their Concorde's for a flight from CDG to New York. The flight (4590) crew began their preflight work, during which they requested the entire length of Runway26 R from Air Traffic Control (ATC) for their takeoff, planned for 2:30 that afternoon. Runway 26 R runs nearly due west (27 would be due west) with a length of 13,780 ft., (it is the longest runway available at CDG). As previously described, the Concorde would take off at a high rate of speed and had planned to begin rotating skyward when it reached 198 knots nearly 228 mph.

At 2:42 the tower granted flight 4590 clearance to takeoff. Twenty seconds after beginning its takeoff roll, the Concorde had reached 100 hundred knots, less than 10 s later it achieved V1 (150 knots, 172 mph), the go/no go velocity. V1 represents the velocity above which the plane, pilot, and passengers are committed to getting airborne. It is now traveling too fast, and with too little runway remaining to safely abort the takeoff. A few seconds after achieving V1, the right front wheel of the main landing gear blew up. The Concorde has 2 main landing gear struts (left and right) with 4 tires attached to each main landing gear assembly.

Forty-five seconds after starting its takeoff role, this Concorde was lifting its nose off the runway and rotating the aircraft to complete its takeoff. The delta wings of the Concorde and its lack of flaps meant that sufficient could only be generated by rotating the nose up and increasing the angle of attack. At this same time the control tower informed the pilot that flames were present behind the aircraft. The copilot acknowledged the tower transmission and the flight engineer (the Concorde relied on 3 cockpit personnel, pilot, copilot, and flight engineer) announced the failure of engine 2 (left inboard engine, engines typically are numbered sequentially left outboard to right outboard).

The following events occurred during the next 30 s of powered flight. The engine fire alarm annunciated, and the flight engineer announced he had shutdown engine 2, while a few seconds later someone pulled the engine 2 fire handle, which flooded the engine area with Halon from high-pressure tanks [3], a common fire-extinguishing material during this time period. Despite its worldwide ban in the Montreal protocol (1989) for damage to the ozone layer, Halon has still remained as a fire extinguisher for commercial airliners as a suitable has not been found.

During this time, the copilot let the pilot know that the airspeed was still critically low (200 knots). The pilot then called for the landing gear to be retracted. This should have decreased drag and normally allow the aircraft to continue its acceleration and climb. The controller at CDG confirmed the presence of a large flames behind the aircraft.

As this first 30 s of flight neared, the engine fire alarm sounded again, for 12 s. About 45 s after the aircraft rotated, the engine fire alarm annunciated again, and continued until the end of the flight. The copilot commented that the landing gear had not retracted and continued to announce the aircraft airspeed several times. Almost immediately after that, the Ground Proximity Warning System (GPWS) alarmed several times. The copilot informed air traffic control that they were trying to reach Le Bourget airport, the former primary airport serving Paris, Le Bourget was relegated to business and general aviation after Charles De Gaulle opened as the primary commercial airport serving Paris. Located about 10 miles southwest of CDG, France still uses Le Bourget for the Biennial Paris Air Show. The westward takeoff direction meant that Le Bourget's location relative to CDG (10 miles Southwest) provided the preferred location for any airplane experiencing a severe emergency after leaving Runway 26R.

Within a short time after the GPWS warnings, data recorders indicated that engine 1 lost power. Just seconds later the aircraft crashed into a hotel in the small town of Gonesse. The crash completely destroyed the aircraft as well as the hotel at the crash site. All 109 persons on the airplane (100 passengers, 9 crew) along with 4 persons at the hotel died from the impact.

The Concorde that crashed at Gonesse was Serial No. 3, which had accumulated 11,989 flight hours and 4873 pressurization cycles. This aircraft entered service in October 1979 and its maintenance record was kept up to schedule. The #3 aircraft had an up-to-date maintenance record and was well below its design life, for both the airframe and the engines.

Accident Reconstruction

As previously noted, the aircraft started its takeoff roll at 14 h, 42 min and 32.4 S. It passed through 100 knots without incident at 14 h, 42 min and 54.6 s, as it accelerated for takeoff. The copilot (first officer) announced "four greens" indicating that all 4 engines were operating as planned and presumably developing appropriate thrust for takeoff. A short 20 s later 14 h, 43 min, 03.7 S, the V1 callout was made (150 knots) as the aircraft continued accelerating through its takeoff run.

A short time and distance down the runway is where the root of the disaster began. The No. 2 tire (inboard front tire of left side main gear) ran over a thin metallic strip. Figure 12.1 shows the location and arrangement of the 8 tires used by Concorde's main landing gear.

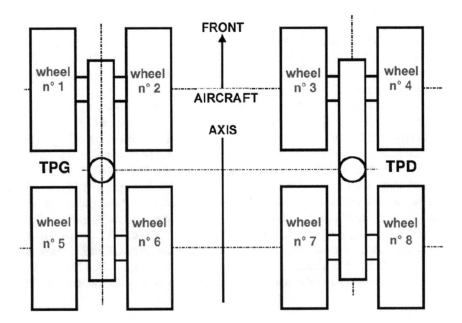

Fig. 12.1 Tire arrangement for Concorde main landing gear

Fig. 12.2 Improper thrust
reverser repair part from
Continental DC-10

Fig. 12.3 Main landing
gear photo with leading
water deflector

The strip that initiated the failure chain has now been attributed to a Continental
Airlines DC 10, falling off that jumbo jet while it was departing CDG, 2 takeoffs
prior to the Concorde. Figure 12.2 shows the strip (~34 cm long by 3 cm wide) as it
was found on the runway after the accident.

A review of the cockpit Flight Data Recorder (FDR) and cockpit voice recorder
(CVR) indicates that thin metallic strip was struck at 14 h, 43 min, 9.5 s. A sharp
noise is heard on the CVR at this time, presumably the No. 2 tire rapidly deflating,
as well as impinging on parts of the landing gear and wing. A slight increase in
rudder load is indicated at this time. The aircraft was traveling at 175 kt (still too low
to rotate and takeoff), and had traveled 1720 m along the runway, 40% of its available length. Postaccident runway inspection found parts of a nonmetallic tire water
deflector, pieces of the No. 2 tire and that metallic strip from Fig. 12.2 at the 1700-m
mark of the runway. Figure 12.3 shows a photo of the main landing gear with the
composite water deflector located just forward of the front tires.

A short 1.5 s later, a distinct difference can be heard in the background noise on the CVR. The Concorde's speed has increased slightly to 178 kt and has used 1810 m of the 26R runway's 4215 m length. The No. 2 wheel started to leave marks of a deflated tire at this same point on the runway. A scant 10 m further down the runway a piece of the bottom surface from the No. 5 fuel tank was discovered and stains from kerosene fuel began to appear. At the 1850-m point, dense soot was noted on the runway, indicating combustion of the fuel. Coupling this information, it seems that a large amount of fuel leaked out of the fuel tank, and then fire broke out and stabilized at the rear of the aircraft behind the No. 1 and No. 2 engines but extending to the fuel source under the left wing (the No. 5 fuel tank). Figure 12.4 shows the location of Concorde's fuel tanks, and the no. 5 tank housed within the port (left hand) wing structure, directly in front of the No. 1 and 2 engines.

One second after the tire burst producing the fuel leak, the captain began to deflect the rudder to the right presumably to compensate for the extra drag on the left side from the burst tire. The aircraft had now traveled 1885 m and reached a speed of 182 kts. Within 2 s of that event (14 h, 43 min, 12 s) engines 1 and 2 suffered their initial loss of thrust, confirmed by the FDR, along with the copilot announcing, "watch out" and the "GO LIGHTS" for both engines extinguishing. The loss of thrust happened so quickly that fire damage seems unlikely, rather that hot gas engine from the flames seems a more plausible explanation for engine 2, while hot gas ingestion or tire debris could have caused the engine 1 degradation.

At nearly the same time (14 h, 43 min, 12.2 s) the captain began to gently pull back on the control column, with the airspeed now 183 kt, and the aircraft nearly halfway down the runway at 1915 m. The thrust was only 50% and came almost entirely from the No. 3 and 4 engines, which generated a large yaw force to the left on the aircraft. The nose gear lifted off the runway just a few tenths of a second later with the aircraft now traveling at 187 kt and passing the 2045-m mark. The low rate

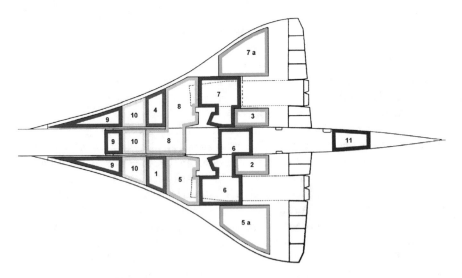

Fig. 12.4 Concorde fuel tank locations

of nose elevation may be indicative of the cockpit crew trying to adapt to the abnormal situation, including a low takeoff speed.

In the span of 3 s the crew had to respond to numerous anomalous data and sensations. Through the CVR they definitely heard unusual noises (the tire explosion, the tank rupture and the subsequent fire), they clearly felt and responded to the lateral accelerations generated by the added drag from the blown tire, they experienced a brief period with all the thrust being generated by the starboard side (the No. 3 and 4 engines), and finally they had to overcome the low airspeed which was affecting their ability to climb away from the airport.

Remarkably, 1 s after these events, the engine 1 "GO Light" re-illuminated, meaning that its fuel flow, pressure rise through its compressor section and rotational speeds approached nominal values, and it began to generate its normal takeoff thrust. Unfortunately, the engine 2 parameters indicated it was generating little thrust, probably akin to an idling engine (3% thrust). A few seconds later (14 h, 43 min, 20;4 s) the flight engineer announced this failure in the No. 2 engine. At this point the aircraft still had its main landing gear on the ground, with a 9° nose-up pitch, was traveling at 203 kt, and had traversed 2745 m of the runway.

Between 14 h, 43 min, 20.9 s and 14 h, 43 min, and 21.9 s, engine 1 suffered a flow surge (likely hot gas or fuel ingestion) which brought its thrust down to the idle level again, generating significant asymmetry as all of the takeoff thrust again came from the right-hand side, the no. 3 and 4 engines.

The increased drag on the left side of the aircraft (deflated tire) coupled with the large difference in thrust between the underpowered left side engines (no. 1 and 2) and full powered right side engines (no. 3 and 4) were forcing the aircraft left despite the pilot's control inputs. The No. 6 wheel (inboard, rear wheel of left main landing gear, see Fig. 12.1) ran over the edge light on the left side of the runway, breaking the light assembly, and indicating that the aircraft had deviated 22.5 from the runway centerline.

Finally, at 14 h, 43 min, 21.9 s, the aircraft completed its takeoff. It was now traveling 205 kt, had traveled 2900 m, and was in a 10° nose-up pitch. One second later the engine fire alarm annunciates. Audible on the CVR are voices from other aircraft waiting to takeoff, identifying the fire on the Concorde, "it's burning and I'm not sure it's coming from the engine." The Concorde's engines used afterburners to help accelerate the aircraft at key points in its flight. Takeoff was one critical point that used this feature, so much like the flames seen on military aircraft using afterburners, it was not uncommon to see flames exiting the Concorde's engines at takeoff. It was unusual to see the magnitude of the flames seen on this day, and they extended well forward of the engine exhaust.

One significant difference between Concorde and subsonic airliners is the absence of a horizontal stabilizer. The tail assembly only had a vertical surface, with the rear section that could pivot to provide yaw control. The pitch up and down control previously provided by the horizontal stabilizer would be delivered by control surfaces at the rear of the main wing. This change was driven by two design requirements. The first was to decrease drag, and any additional control surface added to the Concorde's drag. The second related to the large pitching moment generated by locating horizontal stabilizers at the rear of the tail assembly. Traveling at Mach 2, large changes in the aircraft attitude (up or down) might be too extreme if generated by a tail movement. Instead Concorde moved these control surfaces forward, to the aft end of the main wing. Figure 12.4 shows an overhead view of the tail and wing structure, with the absence of horizontal stabilizers in the tail, instead these features are at the rear of the main wing.

As previously mentioned, the No. 2 engine fire extinguishers were deployed, and about 2 s later the copilot draws attention to the low airspeed (200 kts), when a normal takeoff would be at 220 kt, and a critical speed when operating on 3 engines with the gear still deployed is 205 kts. All aircraft have critical speeds that they need to maintain in the event of an engine failure at critical points in the flight (takeoff being the most critical flight regime for required engine thrust). The flight manual for Concorde indicates that 205 kts is the minimum speed for the aircraft when operating on 3 engines, with the landing gear extended. This number increases to over 300 kts when only 2 engines may be operating with the gear still extended.

Barely 5 s after getting airborne, the copilot drew attention to the very low airspeed (200 kts), when the minimum speed for 3 engines without climbing is 205 kts. The left side engines continued their attempt at recovery, but for engine 1, the ingestion of solid debris meant that it would no longer produce thrust much in excess of idle.

Eight seconds after liftoff, the Captain requested gear retraction. The aircraft was still only traveling at 200 kts, its altimeter indicating a bare 100 ft., and the calculated rate of climb was 750 ft./min. Engine 1 thrust had increased to nearly 75% of nominal, with the afterburner reinitiating. A cabin smoke detector was heard inside the cockpit, whose most likely source was the air-conditioning system used air from the left engines. Despite the air-conditioning packs using compressed "bleed" air from in-front of the combustors and turbines, this engine had ingested a burnt mixture and passed it through to the cabin air-conditioning system.

Seventeen seconds after leaving CDG, the Captain again ordered "gear retraction." 23 seconds into flight, the copilot responds to the Captain's request with "I'm trying." 27 seconds after getting airborne, the copilot reminds the pilot of "the airspeed," with him repeating it 10 seconds later. Thirty-five seconds into the flight, the copilot notes that "the gear isn't retracting."

Thirty-seven seconds after takeoff, the GPWS sounded several times. The copilot communicated to ATC that they were trying for Le Bourget. Engine 1 lost power again, and the aircraft crashed into a hotel at Gonesse. Figure 12.5 shows a photo of the Concorde leaving CDG with significant flames evident below and behind the port wing. This photo shows the large extent of the flames, that the aircraft has started rolling to the left and that the landing gear remained in the deployed state.

Root Cause Analysis

The Concorde during its takeoff acceleration ran over a thin metal strip, aligned vertically, which produced a burst tire releasing significant pieces of tire in both volume and mass. Postaccident tests recreated this scenario at similar takeoff speeds

Fig. 12.5 Concorde with flames leaving CDG July 25, 2000

generating comparable tire remnants, very similar to those found on the runway after the accident.

The tire burst event that started the chain reaction and led to the crash at CDG was not the first event involving a tire on the Concorde. Prior to this flight, the aircraft had experienced 57 different cases of burst/deflated tires. Of these events, twelve had structural consequences, and six of these led to penetration of the fuel tanks. Nineteen (1/3) of the bursts/deflations had been caused by foreign objects on the runway. It seems that the airline operators and the multiple design companies responsible for Concorde were not proactive in understanding the magnitude of the problem that a burst tire could generate.

It was generally agreed that running over the runway debris ruptured the No. 2 tire, an energetic burst that sent large tire fragments into the lower wing, including the No. 5 fuel tank. However, the size of the fuel tank rupture had never been seen on a civil aircraft prior to the Concorde accident. The most likely scenario involved the impact of a large tire fragment at relatively low velocities (120 m/s) striking the fuel tank. This impact then setup a hydrodynamic pressure surge, which breached the tank and allowed fuel to leak out to the environment. The No. 5 fuel tank was full when it was struck by the tire fragment, this impact deformed the tank, displaced fuel and setup a pressure wave that caused the tank to rupture. Postaccident runway inspection found a large part of the No. 5 fuel tank, without impact damage, but was forced out from high internal pressures (the hydrodynamic surge after the tire impact).

Multiple pathways exist for igniting the leaking fuel from the No. 5 tank. However, investigators narrowed their investigation to 2 leading sources. The first and most likely scenario ignites fuel vapor after contacting hot parts of the engine, or the afterburner section. A second but less likely ignition source could arise from arcing of the 115 V cables in the landing gear bay. This second scenario assumes that the same tire damage that ruptured the fuel tank, coincidentally affected electrical harnesses in the landing gear bay area. The landing gear strut and bay doors however provided screening from the tire location, making this pathway less likely than that from fuel vapor contacting hot engine components.

Despite repeated attempts from the cockpit, the main landing gear could not be retracted. There are two hydraulic systems available to perform the gear retraction movement, however both systems need fully opened bay doors in order to initiate the gear retraction sequence, which typically takes 12 s to complete. The main impediment seems to be the left landing gear door, which indicated that it was not in the fully open position during the time that gear retraction was commanded by the copilot. Whether it was a failed sensor, shorted electrical line, or damaged door, or some combination, this failure prevented the retraction of the main landing gear. While keeping the main gear extended increased the difficulty in keeping the Concorde airborne on 3 engines, the large flames under and at the rear of the port wing meant that the flight was doomed. The flames incapacitated both port engines (no. 1 and no. 2) while also structurally weakening the wing and control surfaces. The low-velocity aircraft was extremely difficult to control, only exacerbated by the large asymmetry in thrust (all from the right-hand side).

The French bureau responsible for transport accidents (BEA) and its air accident investigation branch (AAIB) withdrew the airworthiness certificate for Concorde immediately after the accident (Aug 16,2000). The rationale for this strong response came from the realization that a tire failure during taxi, takeoff, or landing is not an improbable event and had generated system damage previously. For this flight, the destruction of the tire resulted in at least one puncture in a fuel tank, resulting in major fuel release. Ignition of the released fuel results in an intense fire for the remainder of flight. Finally, the crew had no means of assessing the fire nor take any action to extinguish it. Therefore, a tire destruction, a simple event that might recur, had catastrophic consequences without the crew being able to recover.

These same authorities defined certain measures that the team that built Concorde would have to implement in order to return this aircraft to service.

1. Installation of flexible linings in fuel tanks 1, 4, 5, 6, 7, and 8
2. Reinforcement of the electrical harnesses in the main landing gear bays
3. Limits to the brake ventilator power supply during critical flight phases, and requiring that the tire pressure detection system be operational
4. Installation of Michelin NZG tires and modification to the anti-skid computer
5. Modification to the non-metallic water deflector and removal of its retention cable
6. Eliminating volatile fuels and an increase in the minimum fuel quantity required for a go-around emergency landing

Introducing flexible linings for tanks 1, 4, 5, 6, 7, and 8 should eliminate the possibility of another tire generating a large rupture in those tanks. The six tanks listed are closest and most in-line with tires from the main landing gear, therefore the most vulnerable to being impacted by a burst tire.

Reinforcing the electrical harness with the main landing gear bay should help limit future accidents via two methods. First, hardened landing gear electrical harnesses will reduce the chance that a landing gear harness could provide an ignition source for any future fuel leak. Second, hardening these harnesses will reduce the chances of having the main landing gear stuck in the open position. As already mentioned, the added drag of the gear, coupled with the low thrust available, and predominantly from 1 side, made the Concorde extremely unstable in flight.

Michelin developed a new radial tire for specifically for the Concorde, after the accident. The Michelin NZG (near zero growth) was designed to handle the weight of a fully loaded Concorde while being much more resistant to damage. The tire relied on new materials and tire reinforcement to prevent the type of bursting seen when their competitors design ran over debris on the runway at CDG [4].

Modifications to the water deflector were intended to limit any opportunity for debris to be thrown into the lower wing/engine inlet. Eliminating the retention cable reduced the mass of material in the wheel area, a key parameter in the amount of damage that could impact the underside of the wing.

The final modification, eliminating volatile fuel, would reduce the propensity for fuel fires, in the event of another fuel tank breach.

All of these modifications were installed on the Concorde and it returned to revenue service with both British Airways and Air France on November 7, 2001, over 15 months after the fatal accident at Gonesse. Early in 2003 both airlines announced that they will discontinue revenue service with the aircraft, ending commercial supersonic flights on Aug 30, 2003. Both airlines cited the increasing maintenance costs of the airliner, coupled with decreasing passenger interest in the premium price commanded by supersonic travel.

References

1. http://mynorthwest.com/824747/forgotten-boeing-sst/? retrieved 8/27/18, Forgotten Boeing SST a powerful symbol of road not taken
2. http://www.pbs.org/wgbh/nova/concorde/anat-nf.html retrieved 8/27/18 Supersonic Dream
3. Davis, "Fire Detection and extinguishing system for Concorde", Aircraft Engineering, Vol 49, No. 10, Oct. 1977
4. http://edition.cnn.com/2001/WORLD/europe/06/07/concorde.tyres/index.html, retrieved 9/18/18, Michelin shows 'safer' Concorde tyres

Chapter 13
Fuel System Failure

TWA Flight 800

The Boeing 747 was the first wide-body commercial airliner. It was Boeing's response to losing the Air Force's C-5 design and build contract for that extra heavy lift cargo plane. But it represented a new approach to carrying passengers in the 1970s (although technically its first test flights were in 1969). This airplane helped to increase the availability of air travel, continue the downward trend in seat-mile costs, and expand passenger service by growing the number of seats available on long domestic and international flights. An iconic airplane was easily recognized by the hump produced by the upper deck located behind the cockpit.

Despite its extraordinary size and position as the most numerous of the initial wide-body airliners, the 747 is a safe plane, at least as measured by aircraft accident statistics. While it was involved in the deadliest plane accident, when two 747s hit each other on a fog shrouded runway at Tenerife on the Canary Islands in March 1977. A KLM flight attempting to takeoff clipped a Pan Am aircraft still on the runway. All 247 passengers and crew on the KLM flight perished while 335 out of the 396 on board the Pan Am flight died. The accident occurred because the KLM flight was confused over directions from the tower and began their takeoff role without clearance. Meanwhile the Pan Am flight had missed their diverting taxiway and was still on the runway [1].

By July of 1997 the 747 type had been involved in ten additional deadly crashes. These accidents were additional to and separate from 2 terrorist attacks which included the June 1985 explosion of an Air India flight near Cork Ireland that killed 329 passengers and crew. Eventually a Sikh extremist was tried and convicted of planning that bombing. The final terror attack was the December 1988 explosion of a Pan Am flight over Lockerbie Scotland by Libyan terrorists that event resulted in 259 killed on the plane, plus an additional 10 fatalities on the ground. In addition, a Korean Air 747 was shot down by a Russian fighter on September 1, 1983, when it drifted into Russian airspace, resulting in the loss of all 269 occupants.

© Springer Nature Switzerland AG 2020
T. Filburn, *Commercial Aviation in the Jet Era and the Systems that Make it Possible*, https://doi.org/10.1007/978-3-030-20111-1_13

Seven of the ten deadly 747 accidents referenced above can be attributed to pilot error. One of the ten was caused by a cargo fire of unknown origin. Because the plane crashed into the ocean due to the cargo fire, it was impossible to retrieve any data on the cause of the fire. However, two of the accidents listed above can be assigned to Boeing's design or maintenance activities. Fortunately, these two accidents occurred on freighter versions of the 747, thereby limiting the loss of life. Both of these accidents occurred when the #3 engine (starboard inboard engine) left the aircraft. These events occurred because the pin holding the pylon to the aircraft wing failed, due to a combination of fatigue and corrosion. In both instances the airflow pushed the engine, into the #4 engine (outboard starboard engine) causing it to leave the aircraft as well. The strong yaw effect (no thrust on right-hand side, all thrust from the left side) combined with loss of hydraulic systems and loss of some control surfaces left the aircraft extremely difficult to control and resulted in crashes. Boeing redesigned the fuse pins that held all the engines onto the wings and no similar incidents have happened [2].

TWA 800

TWA bought and flew 27 Boeing 747s until it ceased operation after bankruptcy in 2001, when its fleet and route network were absorbed by American Airlines. On July 17, 1996, one of those 747s (Boeing 747-100, tail number N93119) began its normally scheduled route taking off at 5:30 am from Athens Greece as flight 881. It landed at New York's JFK airport at 4:30 pm (local times). The Athens crew deplaned after the passengers, but they did not note any anomalies or maintenance squawks from their Greece to NY flight. The new Paris crew boarded and operated the aircrafts tail mounted APU plus two of its three centrally located air-conditioning packs while sitting at the gate for approximately 2 ½ h. During this time a refueling hose was connected to the left (port) wing which when completed left 176,600 lb. of JP-8 in all the tanks before takeoff. The fuel transfer system within the 747 moved the added fuel only among the wing tanks, it did not add of this new fuel to the center wing tank. The only fuel in this center wing tank was that which was left over from the earlier Greece to JFK flight.

The evening weather at JFK was still warm, with the temperature not dropping much from the daytime high of 87 F. A dry day on Long Island became a warm evening, with scattered clouds beginning at 10,000 ft. The temperature was still above 80 F as TWA 800 boarded and got ready to leave the gate at 8 pm. Originally scheduled to leave at 7 pm, one of the ground support vehicles had become disabled and blocked the aircraft from moving. The delay was due to the time required to get a tow vehicle to remove the ground support vehicle. The 747 was finally pushed away from the gate at 8:02 pm. By 8:08 the flight began taxiing to runway 22R (a northeast-southwest runway), where the flight would takeoff on a heading southwest. Not surprisingly this runway headed directly to the Atlantic, minimizing the noise impact of aircraft taking off on local homeowners and businesses.

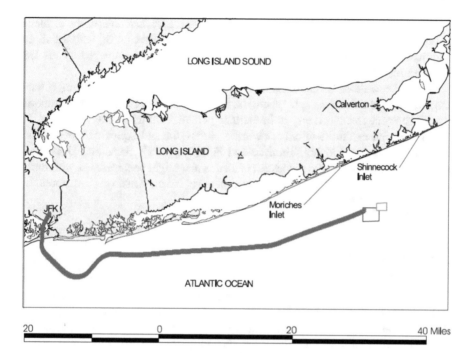

Fig. 13.1 TWA 800 flight path and impact zone [3]

TWA flight 800 was cleared for takeoff at 20:18 (8:18 pm) and became airborne at 2019, with 4 flight crew, 14 cabin crew, and 212 passengers. After takeoff the flight received a series of generally increasing altitude assignments and heading changes from its southwest takeoff to its desired east, northeast flight path for its ultimate Charles De Gaulle destination in Paris. At 20:31 the aircraft exploded approximately 8 miles south of East Moriches, New York. East Moriches is a town located on the south shore of Long Island, 60 miles east of the airport, situated about 2/3 of the island's length closer to its eastern tip. Figure 13.1 shows a map of TWA's flight path and impact zone off the coast of Long Island. In addition to the abrupt stop of the Cockpit Voice Recorder indicative of a sudden event, pilots flying near TWA 800 announced that they saw an explosion near the location of TWA 800. The accident occurred close enough to the coast that citizens on the ground and in boats reported seeing the fireball.

Debris from the accident littered the ocean surface and much of the fuselage quickly sank into the Atlantic Ocean. Most of the debris was eventually recovered and brought to a hangar at the former Grumman Aircraft facility in Calverton NY. The recovery effort used both remotely operated vehicles and divers to retrieve the aircraft debris. The relatively shallow recovery depth meant that the debris field was still located on the continental shelf, the average 120 ft. recovery depth was shallow enough to permit divers to work with air supplied from the surface or with SCUBA equipment. The recovery effort continued for 10 months with contributions from multiple agencies and companies and retrieved more than 95% of the airplane wreckage. An important piece in discovering the cause of the crash, both the flight data recorder (FDR) and cockpit voice recorder (CVR) were salvaged by the US Navy on July 24, 1996.

Many eye witness accounts claimed to see a streak of light resembling a flare moving upward toward the spot where the fireball appeared. Based on these reports the FBI became involved early to investigate potential criminal or terrorist acts. In addition, because of those initial eye witness claims, fire and explosive experts from the FBI, Bureau of Alcohol, Tobacco and Firearms (ATF), FAA, and DoD thoroughly examined recovered pieces for evidence that might be consistent with damage from a bomb, missile, or high-order explosive. No evidence was ever found of any bomb or missile damage.

Accident and Cause

The eye witness accounts of a fireball combined with the wide distribution of wreckage indicated that the flight had experienced a catastrophic in-flight breakup. Once the CVR was analyzed, noises consistent with other airplanes that experienced in-flight structural failures were identified (including those that had undergone fuel tank explosions). These initial facts led investigators to consider 3 possibilities: (1) a center wing tank fuel explosion, (2) structural failure and decompression (see Chap 11), or (3) detonation of a high-energy explosive including a bomb inside the plane or a missile warhead impacting the structure.

Unlike the United cargo door failure (also from a Boeing 747), the forward cargo door from TWA 800 was recovered with all 8 of its latching cams attached to their mating fuselage pins, indicating that the forward cargo door did not depressurize the aircraft. All of the doors were found intact, meaning that none of the doors were involved in the presumed (at that point) explosive depressurization of TWA 800.

Some moderate corrosion was discovered among some of the fuselage's structural members, but no evidence of structural failure was found in these members. These facts led the accident investigation team to conclude that the in-flight breakup of flight 800 was not started by any preexisting structural failure or rapid decompression event.

The original eye witness accounts of a streak of light followed by the accident fireball and led some to believe a missile brought down the aircraft. However, examination of 95% of the airplane did not the find any evidence consistent with a high-energy bomb or missile. No pitting or cratering was found in any of the airplane's structure. Furthermore, none of the victims nor the interior components showed evidence consistent with a high-energy explosion. The missing 5% of the structure were so diverse and spread among such small regions that investigators were confident that no missile nor on-board explosive caused the accident.

Investigators were able to eliminate high-energy explosion and rapid decompression, which left their third potential cause, center wing fuel tank explosion as the

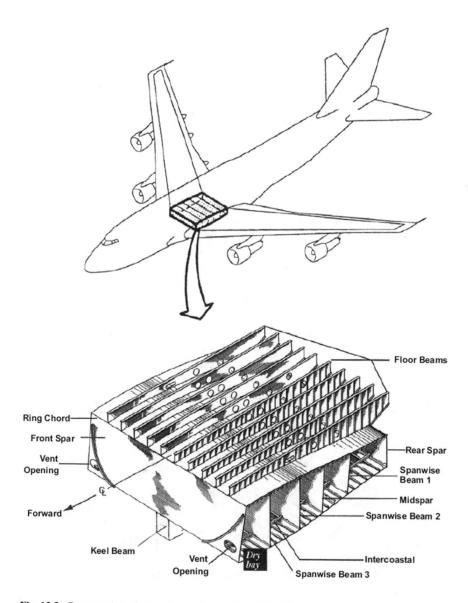

Fig. 13.2 Cross section of Wing Center Section for a 747-100

likely candidate. Strengthening this theory, it became clear while piecing the recovered airplane sections back together that the area around the wing center section (including the center wing tank [CWT]) had been the first sections to leave the airplane. The pieces around the failure initiating point would be expected to depart the aircraft first. Figure 13.2 shows the location of the wing center section on a Boeing747. This portion of the aircraft is about 21 ft. wide, 20 ft. long, and varies between 4 ½ and 6 ft. in height. This section of the airplane supports the fuselage, both wings, and houses the center wing fuel tank which can hold up to 12,890 gallons of jet fuel, nearly 1/4 of the total fuel capacity on this model 747. TWA records indicate that the center wing tank retained 300 lbs. of fuel after landing from its Athens to JFK leg. No additional fuel was added to the center wing tank during the refueling operation at JFK, which would have left the tank primarily full of fuel vapor and only housing about 45 gallons of liquid fuel, less than 1% of its volume capacity.

The air-conditioning packs were operating for over 2 h prior to the flight, and this equipment is located directly below the center wing tank. The heat exchangers, turbines, and compressors that make up the major parts of the air-conditioning systems reject a significant amount of heat to the ambient air surrounding them. This same heat would naturally warm the center wing tank, directly above. The failure investigation team determined that the fuel/air mixture within the CWT was between 101 °F and 127 °F at the time of the accident. The team also determined via testing that Jet A fuel vapors at the same pressure, altitude and fuel loading as TWA flight 800 are flammable down to temperatures of 96 °F. The investigation team also demonstrated that a fuel/air explosion could produce sufficient pressure to break the tank and then destroy the airplane. They tested an out-of-service 747 in 1997 by igniting a propane/air mixture in the CWT. The resulting damage confirmed that the tank failed and would have destroyed a pressurized airplane flying at practically any altitude. Figure 13.3 shows the location and size of the center wing and main fuel tanks on a 747.

Figure 13.4 shows the locations of the three air-conditioning packs for the 747–100. These three units were located directly below the center wing tank. Because

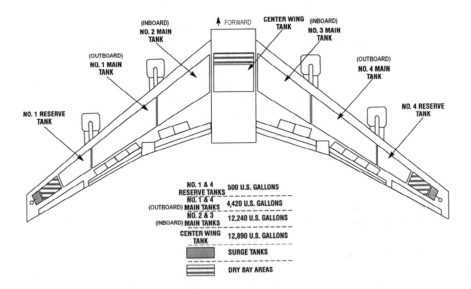

Fig. 13.3 747-100 fuel tank arrangement

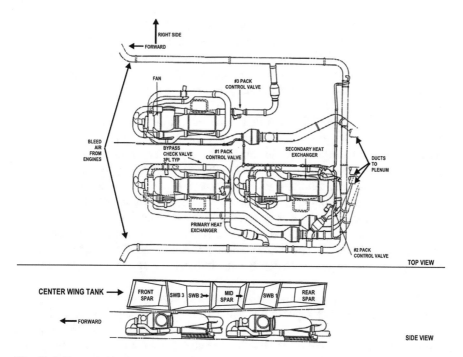

Fig. 13.4 Top and side view of air-conditioning packs below center wing tank

these units rely on high-pressure and high-temperature air from the APU or the main engines, they tend to leak heat into the surrounding structure. As previously noted, 2 of these 3 units were operating for >2 ½ h on APU pressurized air, prior to flight, and presumably switched over to engine bleed air after takeoff for TWA 800.

The metallurgists who examined the wreckage concluded that an overpressure event within the CWT was the initial event in the breakup sequence of this 747. These experts believed that the Spanwise Beam 3 (SWB3, the forward most beam) that separated the CWT from the dry bay in front of it was the earliest piece of the aircraft to be disturbed. The large movement of this beam was clearly evidenced by witness marks (metal to metal high pressure contact) on several adjacent structures. The displacement of this Spanwise beam due to an overpressure in the CWT generated other structural and fuselage skin failures, ultimately but quickly leading to an in-flight breakup of the aircraft.

Postaccident analysis examined a variety of potential ignition sources for the Jet A vapors filling the center wing tank. Boeing's certification concept was based on preventing fuel tank explosions by eliminating all ignition sources. The fuel tank did not include any means to prevent flame propagation nor inerting the atmosphere within the tank. External (to the aircraft) ignition sources had been eliminated by the investigators based on the reconstruction of the accident aircraft which did not display any artifacts or scars that would be expected from a bomb or missile attack.

The two main paths for igniting the vapors in the fuel tank focused on two potential sources, with the first related to the fuel quantity indication system (FQIS). Other ignition sources ranged from autoignition, uncontained engine failure, malfunctioning CWT fuel jettison pump, static electricity, malfunctioning CWT scavenge pump to the failure of a turbine from the air-conditioning pack directly beneath the center wing tank.

Analysis and testing of the Jet A fuel vapor led investigators to discount autoignition due to a hot spot from consideration. A temperature of at least 460 F across a broad surface with hot spots approaching 900 F would have been required to generate autoignition. No evidence was found in the wreckage of any thermal damage to these levels. Even the failure of a high-pressure, high-temperature bleed air tube feeding the air-conditioning packs below the fuel tank would not be expected to generate temperatures as high as those required to autoignite the fuel vapors.

Neither the main engines nor the air-conditioning packs showed any evidence of uncontained failures. Therefore, it is considered very unlikely that either of these rotating components generated debris with sufficient energy to ignite the vapor in the center wing tank.

Investigators considered a malfunctioning fuel jettison pump as a potential spark that could ignite the fuel vapor in the center wing tank. A review of the CVR indicated that the pumps were not used, nor was there any expectation that they had been operated. Even if the pumps had been operated, the combination of the pump location and the flame suppression system inherent in the pump and piping system design make it very unlikely that these pumps were the ignition source.

Several other potential ignition sources were evaluated and similarly dismissed. They included lightning and meteorite strikes, static electricity buildup or fire migration through the fuel vent system. None of these ignition sources seemed credible by the investigation team, based on the evidence available with the reconstructed aircraft.

The only electrical wiring located inside the CWT is the wiring for the fuel quantity indication system (FQIS). Boeing has designed this FQIS to be a low-voltage and low-energy system, which cannot discharge energy in excess of 0.02 mJ. In order to ignite the fuel air mixture in the CWT, a higher than intended voltage would have to be present and that energy would have to be released within the CWT.

The postaccident inspection of the recovered airplane wiring found many instances of cracked or damaged insulation, often enough to expose the underlying conductor. While some of this damage occurred as a direct result of the explosion, crash and recovery operation, an inspection of similarly aged aircraft found similarly degraded wiring systems. This comparison strongly suggests that at least some of the damaged wiring insulation existed before the accident. Damaged wire insulation could lead to electric arcing or short-circuits, an unintended source of energy inside the CWT.

Locations of definitive electrical arcing were found after the aircraft had been recovered. Some of these events may have been caused by the accident, but their number and location suggest that some of these could have contributed to an arcing event within the CWT, through the FQIS. Replay of the CVR demonstrates electric anomalies just before (~1 s) the accident. Finally, the pilot is heard to comment about the "crazy" No. 4 engine fuel flow indicator just 2 ½ min before the CVR lost power and the explosion occurred.

Arcing from exposed conductors for FQIS wiring could provide a mechanism for ignition of the flammable fuel/air mixture in the tank. This scenario was suspected as the ignition mechanism for a 1990 CWT explosion that destroyed a Philippine Airlines Boeing 737 on the ground at the airport in Manila. The explosion led to the complete destruction of the plane in under 4 min. Thankfully most (112) of the 120 passengers and crew escaped. In 1972 a FQIS was responsible for an explosion in an outer wing tank of a US Navy C-130. The pilot was able to land the plane (in a corn field) but the whole aircraft was then consumed by the fire, although the entire crew did safely exit the aircraft.

The fundamental fire triangle requires a flammable gas, oxidizer (typically oxygen), and an ignition source to generate an explosion. In the past Aircraft OEMs and the FAA have assumed that a flammable mixture will **always** be present in aircraft fuel tanks. This assumption comes from the volatile nature of the fuel generating fuel vapor, the gas phase venting requirement, and the accessibility of air (with 21% oxygen) entering through this vent system. They have attempted to preclude explosions by eliminating ignition sources. The loss of TWA 800 and the two other fuel tank explosions listed indicate that designing an aircraft to have no ignition source may not be sufficient to prevent fuel tank explosions. The need to change this design philosophy may be further heightened when one considers the effect of aging aircraft and the damage that inherently accrues on wiring, components and systems throughout their use and the maintenance wear and tear of an in-service aircraft.

Conclusion

TWA 800 was flown by a properly trained and qualified flight crew. The aircraft itself was also properly certificated, equipped, maintained, and dispatched. The in-flight breakup of TWA 800 was not initiated by a preexisting condition (crack or failed mechanical member) that produced a structural failure and then explosive decompression. The aircraft was not brought down by a bomb or missile strike (despite the conspiracy theories that remain to this day). Witness accounts of a streak of light were not related to a missile, instead they represented the burning fuel after the accident and breakup had initiated.

The in-flight breakup of TWA 800 was started by a fuel/air explosion in its center wing tank. External sources (to the airplane) such as lightning or a meteorite strike are unlikely ignition sources for the explosion. Additionally, electromagnetic interference from personal electronic devices or from the aircraft played no role in the accident.

The Boeing design philosophy approved by the FAA that relied on eliminating credible ignition sources while acknowledging the flammability of fuel/air mixtures is flawed. The ignition source for the CWT explosion was most likely the FQIS wiring system. Operating transport-category airplanes with flammable fuel/air mixtures in fuel tanks presents an avoidable risk of explosions. Regulators have focused their highest priority on the CWT as these tanks tend to absorb heat from nearby aircraft systems. The wing fuel tanks tend to colder temperatures and lower explosion risks because of their distance from aircraft equipment and their greater thermal exchange with the surrounding air.

Regulatory Changes

After the TWA 800 accident the FAA started a Fuel Tank Inerting task group as part of their overall Fuel Tank Harmonization Working Group. The Working group was formed to "prevent fuel tank explosions." This Inerting task group was designed to provide a feasibility analysis of fuel tank inerting systems, as a means of preventing these explosions.

The Inerting task group looked at a variety of technologies (including those in use by military platforms) and produced a cost estimate for implementing those that seemed to have a high Technology Readiness Level (TRL).

The Task Group generated the following cost estimate for the systems that could be implemented in the near term (Table 13.1) [4].

Liquid nitrogen is seen as a volumetrically compact fluid to be used for gas tank inerting. By storing the nitrogen as a liquid (its most dense form) it will require the smallest tank volume as well as thinner tank walls for this lower pressure system. The USAF uses liquid nitrogen on-board the C-5 transport aircraft to provide fuel tank inerting. The higher cost from this system presumably comes from the higher cost of producing liquid nitrogen, the specialized cryogenic containers required to retain it, and the lack of liquid nitrogen infrastructure close to airports.

On-board gaseous nitrogen can be cheaper vs. the liquid nitrogen option but adds significant system weight and volume within the aircraft. While it is a mature technology, it was seen as a very undesirable outcome from the operator's perspective because of the weight and volume it would add to the aircraft.

Table 13.1 Cost/benefit for fuel tank inerting systems

Technology	Effectiveness (%)	Cost over 10 years (US dollars)
On-board liquid nitrogen for all tanks	100	$35.7B
On-board gaseous nitrogen for all tanks	100	$33.9B
Air separator module for all tanks	100	$37.2B
Air separator module for center tank	100	$32.6B
Ground-based ullage washing with natural fuel cooling for center tank	99	$4B with gaseous nitrogen $3B with liquid nitrogen

Air separation modules could be chosen from several technologies (including membrane bundles) to provide oxygen depleted air streams for fuel tank inerting. The membrane bundle can take pressurized air (bleed air from the engine compressors, much like the air-conditioning systems) and pass it through a membrane bundle. The oxygen passes through the membrane wall much quicker than the nitrogen, leaving the gas exiting the membrane tube to be oxygen depleted. This nitrogen enriched air can be used to keep the oxygen concentration below 12% in the fuel tank. The oxygen enriched gas that passed through the membrane wall is simply sent overboard. These air separation modules can be used to keep the CWT inert or the entire fuel system, with keeping the CWT inert the cheaper option.

Finally, another technology looked at "washing" the fuel to remove the dissolve oxygen, while a related concept provided for fuel tank inerting while on the ground. These concepts provided a marked reduction in cost, but their Achilles heel came from their inability to provide 100% coverage for fuel tank safety.

The ultimate industry response has been to use membrane modules that create a nitrogen enriched gas stream to be sent to the CWT. The FAA published their Reduction of Fuel Tank Flammability in Transport Airplanes, final rule. This rule required CWT modifications to prevent explosions. Inert gas generating systems that provided this gas in the ullage space over the tank would satisfy this requirement. All new aircraft delivered after 12/27/2010 will need to be compliant. In addition, all large transport airplanes made by Boeing and Airbus that are not cargo only will need to be retrofit by 12/26/2017.

Ultimately the vast majority of the new commercial aircraft manufactured or retrofitted have adopted the Air Separation Module with membranes to generate nitrogen enriched gas streams to inert their CWT.

References

1. Aviation-safety.net/database/record.php?id=19850623-2 retrieved 10/23/18
2. Nederlands Aviation Safety Board, AAR 92011 El Al 1962, Oct. 4, 1992
3. NTSB/ AAR-00/03 TWA Flight 800, Adopted 8/23/2000
4. Aviation Rulemaking Advisory Committee, Fuel Tank Inerting Task Group 3, Report June 28, 1998

Chapter 14
Flight System Sensor Failure

The Airbus A330 is a wide-body (dual aisle) twin-engine aircraft that had its maiden flight in 1992, with Airbus now introducing two new variants (-200 and -300). The plane was conceived as a direct competitor to Boeing's 767 offering. It suffered an auspicious beginning when an A330 on a test flight out of the main Airbus assembly plant in Toulouse crashed as a result of pilot error and inadequate operating procedures. The aircraft entered revenue service with the domestic French carrier, Air Inter in 1994. After the fatal test flight accident only one other event involved a loss of life over the next 15 years of airline revenue service. That lone fatality happened with some very odd actions. A lone Philippine hijacker first attempted to commander the flight (Philippine Airlines flight 812) between Davao City (Philippines) and Manila. After his request to return to Davao City was denied due to lack of fuel, he then elected to parachute from the plane and died when his self-assembled parachute did not operate.

Air France operated revenue transport service between Rio de Janeiro's Galeao airport (Brazil) and Paris Charles de Gaulle. Sunday May 31, 2009, was late fall in this southern hemisphere town, but this city enjoyed a tropical climate, which meant on average a daytime high of 78 °F and a night time low of 64 °F. Its coastal location combined with daytime heating of the ocean water led to frequent clouds in the region.

The Airbus A330 was scheduled to depart Rio de Janeiro's airport (GIG in the 3-letter airport parlance) at 22:00. The ATC cleared the crew to start up their engines and leave the gate at 22:10 local time. The flight was wheels up at 22:29 with 3 flight crew (1 extra to provide relief for the 11-h long flight to CDG), with 9 flight attendants for the 216 passengers. The normally most dangerous part of the trip (combined with landing), the takeoff, was routine with no anomalies.

The cockpit voice recorder (CVR) does not need to store the entire flight's voice data. The FAA and international transport aircraft regulators (e.g., the UN sponsored International Civil Aviation Organization, ICAO) have mandated changes over the years in the requirements for the CVR. Originally designed to hold 30 min

© Springer Nature Switzerland AG 2020
T. Filburn, *Commercial Aviation in the Jet Era and the Systems that Make it Possible*, https://doi.org/10.1007/978-3-030-20111-1_14

of cockpit voice exchange, on magnetic tape media, the newer CVRs hold 2 h of data on solid state recording media.

Reviewing the flight of Air France flight 447 begins after midnight on June 1, 2009. We presume that the takeoff and climb out were uneventful, but the 2 h time limit on the CVR means that we have no data from those events. At this time the plane was cruising at 35,000 ft. (flight level 350) with the autopilot and autothrust engaged. The plane was on a North-Northeast heading for the Southern Atlantic Ocean. At about 00:30 the flight crew received information from the Air France Operations Control Center about a convective region linked to the Inter-tropical Convergence Zone (ITCZ). The Inter-tropical Convergence Zone is that region near the equator where the trade winds of the Northern and Southern Hemisphere come together. This convergence can create frequent and vigorous thunderstorms over land and sometimes the ocean. By 01:00 am the flight was crossing the coast of Brazil heading over the water of the South Atlantic Ocean [1].

Shortly after 01:35 the copilot reduced the scale on the Navigation Display (ND) from the original 320 nautical miles (NM) coverage, down to 160 NM. This reduced scale provided greater definition, and the copilot is heard referring to "a thing straight ahead." After the Captain confirmed this assumed weather issue on the ND, the crew again discussed the fact that the high outside air temperature coupled with the large remaining fuel load meant that the aircraft could not climb to flight level 370 (37,000 ft.).

At 01:45 the airplane entered a slightly turbulent zone, a tropical storm, over the open water as they neared the end of ground radar coverage. The crew dimmed the interior cockpit lighting and switched on the external lights "to see." The copilot noted that they were "entering a cloud layer," and that it would have been good to be able to climb above the weather. A few minutes later, the turbulence increased slightly.

At around 2:00 am, after leaving his Captain's seat, the captain had a briefing with the two copilots who would occupy the cockpit, with the pilot flying being in the right-hand seat (normally occupied by the copilot). The briefing included the Captain noting that "the little bit of turbulence that you just saw we should find the same ahead we're in the cloud layer unfortunately we can't climb much for the moment because the temperature is falling more slowly than forecast," then the Captain left the cockpit. The outside air temperature was a critical parameter for the airplane with its large fuel load. Cold air is denser and can provide greater lift for the same aircraft velocity, warm air simply wouldn't provide enough lift for the aircraft which was moving at Mach 0.82 (550 mph) with a nose-up pitch of 2.5°, typical for this aircraft during cruise.

At 2:06 am, the Pilot Flying (PF, really the younger, more inexperienced of the two copilots in the flight crew) called the cabin crew, telling them that "in two minutes we ought to be in an area where it will start moving about a bit more than now you'll have to watch out there" and finished with "I'll call you when we're out of it." The two copilots discussed the unusually high outside temperatures. The flight crew reduced thrust to achieve a lower speed of M = 0.8 (537 mph) and after that they turned on the engine deicing system.

At 2:10:05 the autopilot disconnected followed by the autothrust system disconnecting, the PF (younger copilot) announced "I have the controls." The left-hand primary flight display (closest to the PF) showed a sharp fall in the *indicated* forward velocity from 315 mph to 69 mph. The suddenness of the change did not seem realistic based on the aircraft flight characteristics; however this abrupt change did initiate the autopilot disconnect, and the switchover to Alternate Law control scheme. The airplane began to roll to the right and the PF made a nose-up and left flight control input. The stall warning alarm triggered twice in a row. This feature (stall warning) alerts the pilots that the lift being generated is dangerously low. Lift generally increases with angle of attack (AOA) but will drop dramatically when the airplane reaches a stall condition. While lift increases with AOA, drag increases too, so the aircraft velocity drops with increasing AOA. A stall can occur when flying too slow, or when climbing at too high an angle, or climbing at a low velocity. The best response to the stall warning, and to avoid entering a stall, is to put the airplane in a nose-down attitude. This response will increase the velocity and decrease the angle of attack on the wings. Both items (increased velocity and decreased AOA) will decrease the chance of stall.

At 2:10:16, the Pilot Not Flying (PNF, again the more senior copilot) said "*we've lost the speeds*" then "*alternate law protections.*" The A330 like all new Airbus aircraft use a fly-by-wire (FBW) linked to a sidestick controller in the cockpit. This control system, first introduced in the A320 aircraft, defines a significant difference from the previous traditional pilot input setup, which used the pedestal/yoke design that traces its lineage to WWI aircraft. The new Airbus FBW system, while more efficient for aircraft weight, meant that the pilot no longer felt the actual response of their movement into the control surfaces nor did the pilot get feedback from the aerodynamic loads on those surfaces. In "Normal Law," the Airbus FBW system had been designed so that the sidestick deflection is proportional to the load applied to the control surface (greater deflection, greater force applied to the control surface). "Alternate Law" meant that proportionality between sidestick deflection and control surface load went away, as well as the control system logic that prevented an airplane stall. **This airplane could be stalled under the "Alternate Law" control software now operating for the A330**. When the flight control system was in "Normal Law" (the vast majority of the time), the computers worked to prevent stall and essentially kept the aircraft from ever entering a stall. The "Alternate Law" version of the control system became enacted on this flight because the speed sensors suddenly stopped providing reasonable input. The "Alternate Law" option had been installed for many scenarios, including one with the speed sensors becoming unreliable [2].

Between 2:10:18 and 2:10:25 (only 7 s), the pilot non-flying (PNF) read out messages from the electronic centralized aircraft monitoring (ECAM) in a haphazard fashion. He mentioned the loss of autothrust and that the flight controls had reconfigured to Alternate Law. He also announced and turned on the wing anti-icing system. Finally, the PNF said that the aircraft was climbing and asked the PF several times to descend. The PF eventually made several nose-down inputs that produced a reduction in the pitch up attitude and the vertical (climbing) rate. However, the airplane was now at 37,000 ft., and continued to climb.

At about 2:10:36 the left side speed display became valid again and read 256 mph. The aircraft had lost over 200 mph in airspeed since they initiated their initial thrust reduction, autopilot disconnect and the initiation of its climb. The cockpit crew had lost a valid speed display for only about 30 s.

At 2:10:47 a scant 40+ s after the speed display was lost and the autopilot disconnected, the thrust controls were pulled back to 2/3 of the IDLE/CLB position (~85% of the engine takeoff thrust at sea level, but significantly less at the altitude they have now achieved). Now the airplane had reduced pitch to 6° with the angle of attack slightly lower, 5°. This setting should provide sufficient thrust for a slight climb while approaching cruise. Unfortunately, the nose-up attitude of the aircraft meant that its climbing velocity (7000 ft./min) was much higher than normal and this thrust setting would be insufficient to maintain their present forward velocity. The aircraft continues to gain altitude (lower air density, lower lift) while also losing forward velocity which magnifies the loss of lift (aircraft lift is a function of aircraft velocity squared, so a 50% drop in velocity equates to a 75% drop in lift).

Starting at 2:10:50 the PNF called for the Captain several times. One second later the stall warning annunciated again, in a continuous manner. The thrust levers were moved to the TO/GA (Takeoff Go-around) setting which would command a higher thrust from the engines. Unfortunately, the FDR does not record who moved any throttle setting, so it is unclear which of the flight crew initiated this change. While the PF made nose-up inputs into the sidestick controller, the stall warning did not change his effort at pulling back on the sidestick controller, which continue to increase the nose-up attitude. The AOA had now risen to 6°, and ominously continued to rise. At this point the 2nd (right hand) speed display also became valid, with all the indicators reading a consistent 213 mph airspeed, well below their cruise velocity of M = 0.82 (550 mph).

At 2:11:06 the airplane reached its zenith, about 38,000 ft. altitude. It has now "bled" so much airspeed that its forward velocity is now down to 213 mph, but its pitch angle and AOA are both 16°. For some reason the PF continues to command a nose-up attitude, despite the stall warnings, low airspeed, and extreme aircraft altitude.

Finally, at 2:11:37 the PNF said "controls to the left," took over control. Perhaps the most damning act in this accident scenario, the PF immediately took back priority without any verbal notification and continued piloting. One major difference between the Airbus sidestick controller and the competing yoke system is the asynchronous implementation of the sidestick system. The yoke system has a linkage so that inputs on one column (either pilot's or copilot's) are translated to pressure and movement in the non-flying column. The sidestick controller system does not have that linkage, they move independently. Any input on one sidestick controller is not transferred to the opposite number.

The Captain returned to the cockpit at 2:11:42, but instead of occupying the left-hand pilot seat or the copilot station, takes a seat behind both copilots. In this position, the captain will have no ability to manually input control but will advise both copilots what to do. The aircraft had reached such a large nose-up attitude that the AOA exceeded 40° and was now in a rapid descent, about 10,000 ft./min.

Over the next 2 ½ min the cockpit crew continued to be baffled by their inability to bring the aircraft into controlled flight. At 2:14:28 the last recorded values indicated a vertical speed of nearly 11,000 ft./min down, and the aircraft still maintained its deadly nose-up attitude, although it had reduced to 16°. No emergency message was transmitted by the crew, and all 228 passengers and crew members perished from the crash.

Accident Investigation

While the aircraft impacted the ocean on June 1, 2009, it was not until June 6 that ships from both the Brazilian and French navies found floating debris. Naturally most of the material floating was low density, such as honeycomb or polymer matrix composite parts. Within days (June 10) the US Navy was assisting the team using passive acoustic searches and towed pinger locators. This initial phase which lasted for ~1 month was not successful in finding the two most important forensic pieces for understanding the accident cause, the CVR and the FDR.

A second phase search by the French oceanographic institute IFREMER used side-scan sonar operated from July 27 through August 17, and it too was not successful in finding the CVR nor the FDR. The following April a third phase was undertaken by a joint Woods Hole Oceanographic and USN team. This team combined visual and sonar searches from both Autonomous Underwater Vehicles (AUVs) with the Triton Remotely Operated Vehicle (ROV) teamed with side-scan sonar from the USN Orion. Again, there was no debris field, and more importantly no discovery of the FDR nor CVR.

Finally, a fourth expedition localized the debris during a search in April 2011. The recovery zone, east of the Mid-Atlantic ridge with rugged terrain features presenting large variations in depth (700 m to 4300 m) made for a very difficult region to pinpoint debris and the recorders. The CVR and FDR were recovered at a depth of 3900 m. Shortly thereafter, a fifth phase resulted in the recovery of both "black boxes," the FDR and CVR, which were brought to the surface on May 1 and May 3, 2011, respectively [3].

As previously mentioned, the two most stressful times of flight, and those with the greatest risk for the aircraft and those on-board are during takeoff and landing. Air France 447 was in cruise ($M = 0.82$) at a comfortable 35,000 ft. altitude.

While there are a multitude of "what if" scenarios, including the flight crew electing to fly slightly higher, or around the thunderstorm activity and its large amount of liquid water content, the precipitating event for this accident was the icing of the pitot probe which provided airspeed data to the cockpit, the flight control computer, and the autopilot system. Figure 14.1 shows a typical location for these probes (multiple including backups) near the nose of the aircraft.

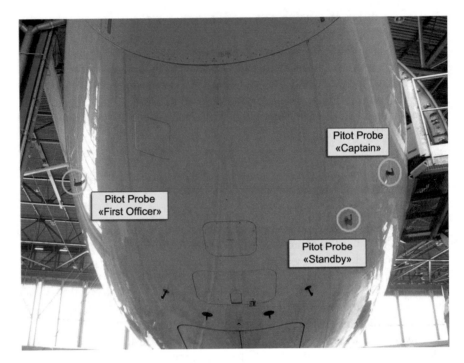

Fig. 14.1 Pitot probe locations on A330

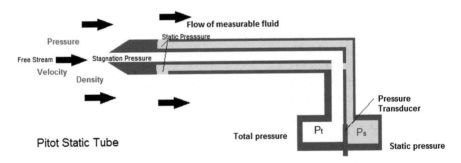

Fig. 14.2 Pitot probe for velocity measurement

The pitot probe operates by measuring the difference in pressure between static unmoving (static) air and the dynamic air pressure developed by the plane's velocity (stagnation) pressure. Figure 14.2 shows a pitot probe with its 2 measurement ports (stagnation and static pressure). The measured pressure difference is a function of the velocity squared. This means that a doubling of the pressure difference will equal a 1.41 increase in velocity (square root of 2 = 1.41). A decrease in pressure difference (which can easily occur if the stagnation pressure port becomes clogged with ice crystals) will erroneously indicate a lower velocity. While almost all (if not

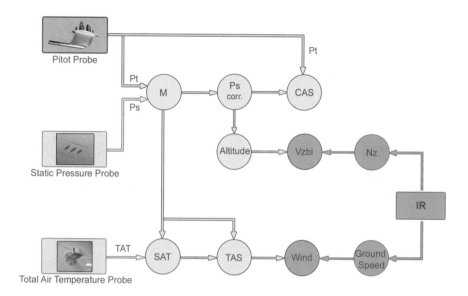

CAS : Calibrated Airspeed - speed *indicated on the PFD*
TAS : True Airspeed - *aircraft velocity relative to the air mass*
M : Mach Number - ratio between true airspeed and sound velocity
Ps : Static Pressure - *pressure of outside air*
Pt : Total Pressure - *static pressure added to the pressure due to aircraft speed*
SAT : Static temperature - *outside air temperature*
TAT : Total Temperature - *static temperature added to the temperature due to aircraft speed*
Vzbi : Baro-Inertial Vertical Speed
Nz : Vertical Load Factor

Fig. 14.3 Pitot probe, static port, and total air temperature probe

all) pitot probes on transport aircraft house heater elements, the magnitude of the supercooled liquid droplets in the thunderstorm clouds that flight 447 flew through apparently overwhelmed this heating element and blocked some or all of the stagnation port.

As shown in Fig. 14.1 multiple pitot probes operate on the aircraft and their output is combined with the total air temperature probe (vital when combined with static pressure measurement for calculating air density) for determining Mach number and the aircraft's calibrated airspeed (CAS). Figure 14.3 shows how these different measuring systems (pitot probe, static pressure probe and total air temperature) are combined to achieve these vital aircraft speed measurements.

It is important to remember that Fig. 14.3 represents the output of one pitot probe, static pressure port, and total air temperature probe. The A330 had 3 different sets of pitot probes (one for the pilot, one for the copilot, and one installed backup). Figure 14.4 shows how the pneumatic lines from these different ports were combined in the integrated standby instrument system (ISIS). In this figure air date reference (ADR) calculates the calibrated air speed and Mach number, while the air

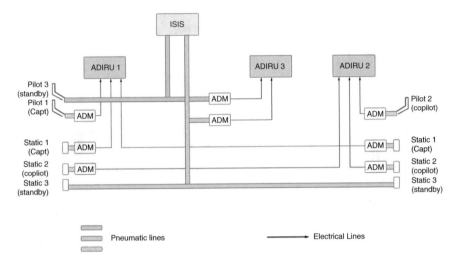

Fig. 14.4 Speed measurement system architecture

data modules convert the pneumatic pressure measurements into electric signals. Finally, the output of the ADRs is evaluated by three different air data inertial reference units (ADIRU, flight computers).

As previously mentioned, the icing and then loss of accurate speed indication from the pitot probe was the initiating event for this accident. However, as in most catastrophic events, there is a failure chain, that had multiple chances to be "broken" and avert the calamity, in this case the loss of 228 lives. The icing pitot probe produced the autopilot disconnect, as the computers had been programmed to do.

Unfortunately, the PF's response to the autopilot disconnect was excessive and ultimately led to a loss of control of the airplane. His initial response, pulling the sidestick controller back, generating a large nose-up attitude, may have been due to the suddenness and surprise of the autopilot disconnect, his continued persistence in this nose-up attitude is not easily understood. Exacerbating his uncommon nose-up response was his lack of a verbal confirmation of this nose-up, sidestick input to the PNF. A clearly contributing event to the eventual loss of control was the lack of understanding by anyone in the cockpit (including the Captain when he returned later) that the computer control scheme was no longer in the normal control mod. Instead, the autopilot dropout meant the aircraft had been released to a more permissive (in terms of pilot input and airplane response) alternate law mode.

As mentioned previously, stall is a large drop in the lift generated by the wings. For large transport aircraft (e.g., A330) it can have two major sources, insufficient forward velocity or an angle of attack (AOA) too steep. The AOA represents the angle between the aircraft velocity vector (the direction it is traveling) and the wings. Figure 14.5 shows this relationship between lift coefficient (non-dimensional representation of the lift force generated by the wings, y-axis) and the AOA (x-axis). It can be clearly seen in this figure that the higher speed that flight 447 was traveling at (M = 08) will reach the stall limit at a much lower AOA, therefore making it easier

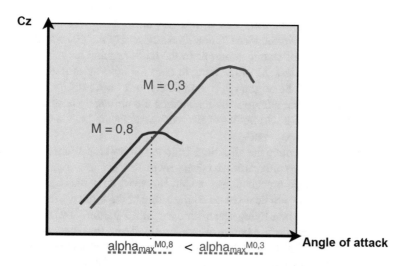

Fig. 14.5 AOA vs. coefficient of lift at 2 velocities

to stall the aircraft and lose control. Higher AOAs create significantly higher drag, even after the stall point. So, the effect of increasing the nose pitch up attitude was to increase the AOA, increase drag, and force the aircraft to climb. All three of these effects from increasing AOA (increased drag, nose-up attitude and higher altitude) combined to greatly reduce the forward velocity, again increasing the probability of entering a stall. The generally preferred response to leave a stall state is to put the nose of the aircraft down, and with gravity and/or the engines increase the airflow velocity over the wings at a lower AOA.

After the second stall warning annunciated, neither the PF nor the PNF responded either orally, nor with any noticeable change in control input. The alarm went off for 9 s at which time it reached its operating ceiling (for its weight at that time) with a relatively high vertical speed, but with a flight path speed that dropped precipitously due to the effects of stalling. As anyone who has launched an object with a sling shot, the aircraft had reached its flight apogee, converting all of its kinetic energy (velocity) into potential (highest altitude). Even though the engines had been placed into a high thrust position (Take off/Go Around, TOGA), the drag due to the high AOA and the fact that the thrust was angled downward as well as aft meant that the aircraft could not accelerate (at least in the direction that the nose was pointed). The PF response to these speed indications and stall warnings led investigators to believe that he was concerned with the aircraft going into an overspeed condition, not a stall. Less than 1 min after this second stall warning he moved the thrust levers down to the IDLE level, and a few seconds later attempted to deploy the speed brakes. Even though the PNF had called out the reconfiguration to "Alternate Law," it seems likely that the PF did not assimilate what that change meant, including that the aircraft could now reach a stall condition based on pilot input.

The PNF's response to this second stall warning was to call for the Captain, which seemed to occupy a large part of his attention. He did not verbalize anything in relation to the stall warning, and it was not until 45 s after the alarm and immediately after the PF announced "I don't have control of the plane at all" that he took the dramatic step of stating "controls to the left" meaning he was now flying the airplane. Unfortunately, he did nothing to alter the nose-up altitude, but made two lateral left inputs. The original PF immediately took back priority and kept his sidestick controller at the left stop and maintained the ultimately destructive nose-up attitude. This action could not be explained by rational experts examining the FDR and the cockpit crew inputs.

The Captain had to notice the vibrations from the aircraft buffeting from its stall condition, as the nose-up pitch attitude now close to 15°, in addition he had to hear the stall warning as he approached the cockpit. However, upon his return, he made no mention of these, nor did he insist on taking either of the flight positions, instead positioning himself in a non-flying seat behind the pilot's station. The stall warnings continued to annunciate, while both copilots informed him that they had lost control. The PNF stated that he had "tried everything"; however neither copilot provided the pilot with a synopsis of their situation, nor the initiating events. When the Captain reentered the cockpit, the altitude was close the cruise level at which he left, which made it difficult to assess the problem. In addition, the rapid descent rate and the stress conveyed by the PNF indicated the urgency and brevity with which the Captain had to implement meaningful control changes. His inadequate interventions meant that he had not identified the stall, despite the continued warnings of the audible stall warning that sounded in the cockpit.

The Captain had only a brief period (seconds) to assess the situation and provide positive input that could have saved the aircraft and passengers. Within 20 s after entering the cockpit, the aircraft was in a 40° nose-up attitude, descending through 31,500 ft. with a descending velocity of about 10,000 ft./min.

The accident was initiated by a pitot probe becoming blocked by ice crystals generating an inaccurate air speed indication. This led to the loss of airspeed information and the disconnect of the autopilot system. The two copilots in the cockpit at that time failed to understand the situation and lost control of the aircraft.

Postaccident Recommendations

The first recommendation put forward by the investigation team did not relate to the accident or its immediate cause. Instead, it related to the 23 months that intervened between the accident and finding both the CVR and FDR. It recommended that EASA and ICAO extend the postaccident transmission time for the underwater locator beacons (ULBs), housed with the FDR, to 90 days. It also recommended that an additional ULB transmitting between 8.5 kHz and 9.5 kHz be added to overwater flights. This new frequency would aid the long distance identification of ULBs while the 90 day battery life would allow searchers more time to locate these very important postaccident items. The team recognized that it can take days up to a week for search ships to reach the site of a mid-ocean crash, which meant this extra time would help searchers trying to locate the FDR and CVR.

The remaining recommendations pertained to the aircraft and how the crew handled the situation. The team recommended that EASA review of cockpit training and make approach to stall, stall recovery and high-altitude stall situations mandatory. Neither of the two copilots had been trained in the approach to stall nor stall recovery at high altitude.

The crew (including the Captain) never formally identified their stall situation. Information on angle of attack (AOA) is not directly accessible to the pilots. For this Airbus aircraft, the AOA during cruise is close to the stall warning trigger AOA when the aircraft is not in normal control law. Under these conditions, pilot input can introduce higher AOA such as that seen in this AF 447 accident. It is vital for flight safety to quickly reduce the AOA when a stall is imminent. Only by having the direct readout in the cockpit of the AOA can flight crews quickly diagnose the proximity to stall and quickly provide input to maintain control of the aircraft. This recommendation asked EASA and the FAA to evaluate the relevance of an AOA indicator directly accessible to the pilots in the cockpit display. The review board made several other recommendations in regard to recording flight data and transmitting position data more frequently to help identify airplane locations in the event of an emergency.

References

1. BEA Final Report, Accident on 1st June, 2009, Air France 447, updated 7/27/2012
2. http://www.popularmechanics.com/technology/aviation/crashes/what-really-happened-aboar-dair-france-447-6611877 retrieved 11/2/18
3. BEA Sea Search Operations Report, Accident on 1st June 2009, AF 447, published Oct. 2012

Chapter 15
Icing Conditions

Halloween 1994 (10/31/94) was an auspicious day for the Chicago area. Typical for the late fall, the day began cloudy with temperatures near 40. American Eagle flight 4184 was an ATR-72 turboprop aircraft that was scheduled for five segments on that day initiating travel from Chicago's O'Hare airport. It had planned stops at Indianapolis IN (IND), back to Chicago IL (ORD), then to Dayton OH (DAY), return to Chicago (ORD), and a final stop at Champaign/Urbana IL (CMI). While a cold driving rain would soak the Chicago and surrounding area, the worst of the weather and its effects happened aloft.

American Eagle Flight 4184 was operating on a Simmons Airline ATR72 aircraft. Simmons airline was doing business as an American Eagle flight. This American Eagle setup was very similar to the arrangements used by the major airlines in order to route smaller feeder aircraft conveniently into their hub locations. This process let them convey more passengers on their larger turbofan aircraft. Simmons airline has since become part of a wholly owned subsidiary to American Airlines still operating under the American Eagle name. However, these commuter airlines can remain autonomous while still doing business under the American Eagle name (Delta Connection or other commuter airline name). Such was the case in 1994 when Simmons was operating that ATR72 aircraft.

The ATR 72 was a product of a joint venture between EADS (European Aeronautic Defence and Space, a major European aerospace company that holds 80% of the Airbus consortium) and Finmeccanica/Alenia Aeronautica, an Italian aerospace company now primarily doing business as Leonardo. The final assembly of the ATR72 occurred in Toulouse, the same site that the ATR72 had its maiden flight in 1988. It began revenue service in October 1989 with Finnish operator Finnair. ATR which has both an Italian and French translation, literally means Air Transport Regional. As a twin-engine turboprop airline, it adopted the high wing (above the fuselage) configuration presumably because of the powerplant (turboprop) and the greater ground clearance for the propellers provided by the high wing. It also incorporated a T-tail design that moved the control surfaces above the wing wake region and allowed them to be more efficient in their control. The ATR

© Springer Nature Switzerland AG 2020
T. Filburn, *Commercial Aviation in the Jet Era and the Systems that Make it Possible*, https://doi.org/10.1007/978-3-030-20111-1_15

Fig. 15.1 ATR72 Aircraft

program originated with the 48 seat ATR 42. The ATR72 can hold 72 passengers with a 29-in. seat pitch. Figure 15.1 shows a photo of the ATR72 aircraft Roselawn accident.

After entering airline service in 1988, the ATR72 had no fatal accidents prior to 1994. Production of the aircraft has continued through 2018 at the Toulouse facility. While it suffered no fatal accidents during its first 6 years of service, it did require an unusual tail stand to be installed after landing. This support rod near the rear of the aircraft prevented the aircraft from striking the runway during loading or unloading operations that inadvertently moved the Cg behind the main landing gear.

The ATR72 that ultimately would become American Eagle Flight 4184 began the day as American flight 4101 with a late morning departure scheduled to arrive in Indianapolis (IND) early in the afternoon. The flight crew reported for duty at 10:39 am (central time) and departed O'Hare (ORD) on schedule at 11:39 am, arriving at IND at 12:42. Using the same aircraft and flight crew and aircraft, the Flight was now 4184 with a 14:10 scheduled departure and a 15:15 planned arrival at ORD. The flight left the gate nearly on-time 14:14, but the deteriorating weather conditions meant that air traffic control (ATC) kept the plane on the ground at IND for an additional 42 min. Similar to many of the shorter flights common for this aircraft type, it was intended to top out at 16,000 ft. altitude. In fact, it was this flight profile that made this turboprop more fuel-efficient vs. a turbofan (jet) engine. A turbofan operates more efficiently at high altitude and high speed (M ≈ 0.8). However, a turboprop operates more efficiently at lower altitudes and lower speeds. Hence the shorter routes, like this 180 mile flight between IND and ORD, which don't have time to reach higher altitude, will save fuel by operating turboprop aircraft.

At 14:53, the local ground controller informed the crew that "you can expect a little bit of holding in the air and you can start 'em up [referring to main engine start] contact the tower when you're ready to go." The crew did not get any info on the reason for their present ground hold, nor why they were to expect an airborne hold [1].

At 14:55 the local ground controller cleared flight 4184 for takeoff, on a planned 45 min flight using Instrument Flight Rules (IFR, vs. Visual Flight Rules, VFR). The crew were to fly directly to an intermediate radio beacon, to a second and then third radio beacon and then to ORD. All of these radio beacons were closely aligned with their North Northwest flight path to Chicago.

Reviewing the Flight Data Recorder (FDR), it indicated that the flight crew engaged the autopilot as the airplane climbed through 1800 ft. At 15:05 the captain reported that they were at 10,700 ft. and climbing to 14,000 ft. At 15:08 the captain requested and received ATC clearance to continue climbing up to the final en route altitude of 16,000 ft.

At 15:13 flight 4184 began to descend to 10,000 ft. altitude in its initial downward slope to ORD. The FDR identified the activation of the Level III deicing system. The level III system activated the wing, horizontal and vertical stabilizer leading edge "boots." These boots were the typical pneumatic deicing boots popular on smaller commuter and General Aviation (GA) aircraft. As detailed in Chap. 8, these pneumatic "boots" inflated periodically (1/min) to shed ice from the surfaces they were attached to.

At 15:17 the ATC office controlling the flight through Indiana was advised by O'Hare Traffic Control to issue holding instructions to inbound aircraft. One minute later the same Indiana ATC put the ATR72 in a holding pattern at 10,000 ft., with an anticipated clearance to land about 15:30 local time. The holding pattern consisted of a 10 nautical mile legs, coupled with a speed reduction. At 1524 the captain of flight 4184 reported "entering the hold" to the Indiana ATC. A new update from the automated flight deck indicated the hold was likely to be in effect until 15:45. The flight was still at 10,000 ft. altitude but had reduced speed to 175 knots indicated air speed (KIAS). The airframe deicing system was deactivated and the propeller system was reduced to 77 percent. The AOA was approximately 5° at this time with the flaps completely stowed. Ten minutes later while in the holding pattern the first officer moved the flaps to the 15° position, in order to improve the handling of the aircraft at the lower speed of the holding pattern.

At 15:38 the Indiana ATC provided a new expected further clearance (EFC, authorization to begin the landing sequence) of 16:00 for flight 4184. At 15:41 the FDR recorded the activation of the level III airframe deicing system, again. The propeller speed increased to 86 percent, the minimum required speed to dislodge propeller ice while in this state. The propeller deicing system did not cover the entire blade, instead it melted some of the ice near the root and relied on centripetal force to sling the ice off the remainder of the blade, but this sufficient force could only be generated when the blades operated at 86% or higher rotating speed.

At 15:48 the CVR recorded one of the two pilots saying, "I'm showing some ice now," but could not discern which flight crew member spoke. At 15:56 the first officer commented "we still got ice" but did not receive an acknowledgement from the Captain. Twenty seconds later the Indiana ATC instructed the flight crew to "descend and maintain eight thousand [feet]." This transmission was followed by one informing them that "it should be about 10 min till you're cleared in."

At 15:57 as the airplane was descending through 9130 ft. on its way to 8000, the AOA increased through 5°, and the ailerons (control surfaces at rear of the wing) began deflecting to a right-wing-down (RWD) position. Less than 1 s later, the ailerons quickly deflected to 13.4° (out of 14° design max) RWD, and the autopilot disconnected. The airplane quickly rolled to the right (not surprisingly based on the aileron input) and the pitch attitude and AOA began to decrease.

Within a few seconds the AOA dropped through 3.5°, the ailerons moved to a nearly neutral position, and the airplane stopped its rolling motion, but at 77° RWD. The airplane started to roll toward a wings-level attitude, the elevator began moving to achieve a nose-up attitude, and the AOA began increasing with the pitch attitude stopping at about 15° nose down.

Five seconds after the initial (and completely surprising) roll to the right, and while the aircraft was rolling back to the left (now at 59° RWD), the AOA increased again through 5° and the ailerons once again deflected to an RWD position. The FDR recorded that the Captain's nose-up control column force exceeded 22 lb$_f$ while the right rolling rate exceeded a stomach churning 50°/s! In fact, during this event, the airplane went inverted (180°) and completed this roll to a wing-level attitude. The nose-up elevator command and the AOA decreased rapidly and the ailerons now deflected to a 6° LWD, but then moved again to a 1° RWD position, and after completing its first roll, the airplane stopped rolling at 144° right wing down.

Three seconds after the first complete roll to the right, the airplane began tolling to the left and the airspeed increased to 260 knots, while the pitch attitude remained nose down fluctuating up to 60° down, and the altitude decreased down to 6000 ft. A scant 3 s later the roll attitude passed through 90° RWD going to wings-level, but the pitch attitude had grown to 73° nose down, and the airspeed was now 300 KIAS, and the altitude had decreased to 4900 ft. During these flight maneuvers (both commanded and uncontrolled) the aircraft experienced accelerations between 2 and 2.5 G.

As the pilot and first officer struggled to regain control, the acceleration had increased to over 3 G's and the Ground Proximity Warning System (GPWS) can be heard in the CVR alerting the flight crew about their low altitude. Less than 2 s later, as the aircraft sailed through 1700 ft. altitude, the first officer can be heard making an expletive comment, also on the CVR. The final recorded info on the FDR shows an altitude of 1.62 ft., 500 ft./s vertical downward velocity, 375 KIAS, a nose-down pitch of 38° and the elevators positioned to achieve 5° nose-up attitude. The aircraft hit the ground in a wet soybean field in Roselawn, IN. The plane was partially inverted, in a nose-down, left-wing-low orientation. There were no survivors among the 64 passengers, 2 flight crew, and 2 cabin crew members.

Accident Investigation

The crash investigation and its evidence revealed that the flight crew on board American Flight 4184 experienced a sudden autopilot disconnect produced by an uncommanded and large aileron deflection, producing a rapid roll of the airplane. The roll response of the airplane was consistent with airflow separation near the ailerons, and generated a ridge of ice that formed aft of the pneumatic deicing boots, on the wing upper surface.

The FDR indicated that prior to the aileron upset that produced the accident, at least 2 instances of drag increase were noted. These occasions of increased drag were consistent with ice buildup occurring while the aircraft was in its holding pattern waiting for clearance to land at ORD. The first occurred at 15:33 just before the flaps were deployed to 15°, the second at 15:51 about 6 min before the fatal upset.

Based on the FDR data of increasing drag and the known meteorological conditions, it is likely that the airplane was intermittently entering areas of supercooled drizzle and rain drops while in its holding pattern. This liquid generated ice on the upper surface of the wing, including regions behind the pneumatic deicing boots, but in front of the ailerons. The wings and horizontal stabilizers use pneumatic deicing boots (see Chap 8), while the propeller, windshields, and engine air intakes all use electric heat deicing. Figure 15.2 shows the various deicing equipment and ice sensors on-board the ATR 72 aircraft.

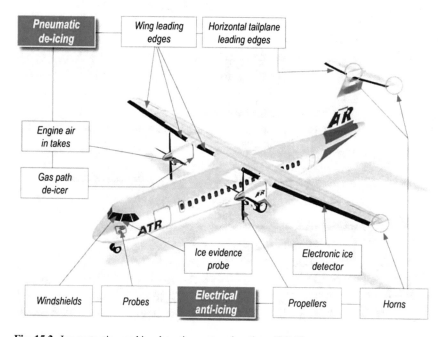

Fig. 15.2 Ice protection and ice detection system location, ATR 72

As the plane began descending, their speed naturally increased, setting off the overspeed warning as they had previously put the flaps to a 15° position. The pilots retracted the flaps and the autopilot put the nose in pitch up orientation to limit their vertical descent rate.

As the plane began that pitch-up maneuver, the AOA increased through 5° and the airflow near the right aileron became separated, presumably because of ice buildup. This flow separation greatly altered the flow pattern, generating a negative pressure field above the right aileron. This flow separation produced a hinge moment reversal on the right aileron, quickly raising it to its maximum position, and generating the rapid RWD attitude previously described.

The pneumatic deicing boots used by both the ATR42 and ATR 72 aircraft protected the leading edge and short distance along the top surface of the aircraft. The pneumatic boots common to both the ATR style commuter and GA aircraft were only installed along the leading edge of the wing. The concept was that breaking up the ice in these positions would generate significant aerodynamic forces along the remaining wing surface to keep it clear of ice. Or if it did not remove this ice, it would be far enough away from the leading edge to not pose a hazard to the wing performance. The effect of the flow separation was to generate a low-pressure air field downstream of the ice buildup. This low-pressure air produced a rotating force on the right wing aileron, which overcame the holding force of the control system. The ATR aileron relied on a series of cables, bellcranks, and push-pull rods to move them from within the cockpit. This system was simple and light but it did not have a hydraulic assist to increase the pilot force on the control surfaces. Figure 15.3 shows the setup of the linkage between the cockpit and the ailerons.

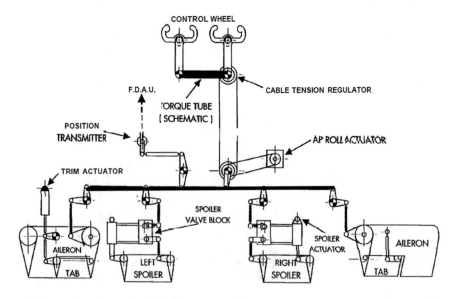

Fig. 15.3 ATR roll control scheme

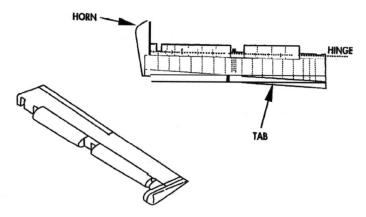

Fig. 15.4 Aileron assembly

The aileron used an exposed horn (outer edge) that extended in front of the hinge line to limit the aerodynamic forces that would be generated by the aileron extending (either up or down) into the flow stream. This horn and the offset hinge line that limited this aerodynamic force can be seen in Fig. 15.4, the Aileron Assembly.

The initial pilot response, LWD aileron, and nose-down elevator deflection reduced the AOA. The aircraft began to roll back toward wings-level. The crew then applied a slight left rudder input and nose-up elevator. As the AOA again increased above 5°, the airflow over the right aileron separated again, resulting in a second aileron hinge moment reversal and a second rapid RWD deflection. It appears that the aircraft again rolled to the right for about 9 s, rolling through at least one complete roll and starting a second. During this time, the AOA remained above 5°, insuring continued flow separation over the right aileron and continued roll control instability. The aircrew now reduced the nose-up input, allowing the AOA to decrease below the 5° AOA, letting the airflow reattach over the right aileron.

Unfortunate for the crew and passengers on Flight 4184, the airplane had now dropped to 6000 ft., at a descent rate of 400 ft./s but had also accelerated to over 250 KIAS, and with a severe nose-down (73°) attitude. The airplane continued descending (3700 ft. altitude), the airspeed increased (327 KIAS) but the pitch attitude improved (60° nose down) but was decreasing. A few seconds later, the Ground Proximity Warning System (GPWS) began sounding "TERRAIN TERRAIN" warnings. Less than 2 s later the aircraft was still descending, now at 1700 ft. The elevator position increased as the crew desperately strove to pull the nose up, but the vertical acceleration increased to 3.7 G, and a loud crunching sound could be heard through the CVR. The Safety Board concluded that the crunching sound came from the large nose-up elevator input coupled with the high airspeed. This combination likely exceeded the designed wing loading and resulted in structural failure of outboard sections of the wings.

The safety board went to the extraordinary step of testing an ATR 72 under in-flight icing conditions. They used a modified USAF KC-135 tanker aircraft with a special water spray nozzle at the end of its refueling boom to produce the icing environment at altitude. The purpose was to examine the upper limits of the icing certification envelope as well icing from very large droplets, not covered by the ATR certification. These tests did not recreate the hinge moment reversal seen by the Flight 4184, but they did demonstrate significant ice buildup downstream of the deicing boots, especially when the flaps were set to the 15° position. The ice buildup did seem to degrade the handling characteristics of the aircraft after the flaps were returned to the 0° position.

The safety board enlisted the help of NASA through a combination of computer simulation and testing at their large icing wind tunnel at the Glenn Research Center in Cleveland. Both the simulations and the testing indicated that ice accretion could occur downstream of the active deicing boots, when operated in conditions like those seen by Flight 4184.

When the safety board reviewed the past history of this aircraft type (ATR 42 & 72), they discovered that 24 roll incidents had been reported since 1986, immediately after it initiated service in Dec. 1985. Of these 2 dozen roll events, 13 were found to be related to icing incidents. Of those 13 icing related, 5 occurred in similar conditions to the freezing drizzle/ freezing rain experienced by Flight 4184 and had varying degrees of uncommanded aileron deflection. All five of these incidents were investigated by either the NTSB, the French equivalent (BEA), or ATR the aircraft builder. The major response to these events was an "Operators Information Message" (OIM) from ATR that indicated the deicing system was not designed for operation in freezing rain (where ice could coat broad sections of the aircraft surface). It also indicated that the icing condition could affect the autopilot and the induce large bank angle changes in the aircraft, eerily similar to the Roselawn accident. In addition to the notice to ATR operators, ATR also made small changes in the aircraft wing design. These changes included a vortex generator in front of the ailerons and an additional Anti-Icing Advisory System (AAS). The vortex generators were intended to raise the AOA at which the ailerons might become unstable. The AAS was an automated system for detecting ice accumulation, but anecdotal pilot information indicated that it rarely warned of ice before the pilot or copilot observed it.

The result of the Roselawn accident investigation ultimately showed a larger rift between the BEA and the NTSB. The BEA took exception to numerous conclusions drawn by the NTSB, to the point that a second report was issued by the NTSB that documented the BEA's response to the NTSB draft report. There were some inflammatory statements in the NTSB report that did not paint ATR (nor the BEA) in a complimentary light. One paragraph of the NTSB final report read:

> "*Based on the long history of ATR incidents in icing conditions, especially those that occurred after 1992, the DGAC (*Directione Generale de L'Aviation Civile, French equivalent to FAA) *should have recognized that the vortex generators, the AAS and the All-Weather Operations brochure were not sufficient to correct or prevent the recurrence of the ice-induced aileron hinge moment reversal problem. Further, it should have been clear that the ATR airplanes were still being flown into icing conditions that were beyond the Appendix C envelope or were otherwise conducive to aileron hinge moment reversals.*"

The more significant NTSB conclusions (again many disputed by DGAC) are summarized below:

1. *If the FAA had acted more positively after the NTSB's aircraft icing recommendation in 1981 (*following a series of aircraft icing-related accidents)*, this accident may not have occurred.*
2. *ATR 42 and 72 ice-induced aileron hinge moment reversals, autopilot disconnects, and rapid, uncommented rolls could occur if the airplanes are operated near freezing temperatures and water droplet median volume diameter (MVD) typical of freezing drizzle.*
3. *The DGAC and the FAA failed to require the manufacturer to provide documentation of known undesirable post-SPS (stall protection system) flight characteristics, which contributed to their failure to identify and correct, or otherwise properly address, the abnormal aileron behavior early in the history of the ATR icing incidents.*
4. *Prior to the Roselawn accident, ATR recognized the reason for the aileron behavior in the previous incidents and determined that ice accumulation behind the deice boots, at an AOA sufficient to cause an airflow separation, would cause the ailerons to become unstable. Therefore, ATR had sufficient basis to modify the airplane and/or provide operators and pilots with adequate, detailed information regarding this phenomenon.*

NTSB Recommendations

The NTSB issued several recommendations for the FAA as well as American Eagle operators using the ATR 42 and 72 aircraft.

For the FAA, the recommendations included:

Revising the icing criteria as listed in 14 CFR part 23 and 25 (transport aircraft icing regulations) to include info from the latest research on ice accretion under varying atmospheric conditions.

Revise the icing certification testing regulations to ensure that all airplane types are properly tested for all conditions in which they are authorized to operation. If safe operation cannot be demonstrated by the manufacturer, they should impose operational limits.

Encourage ATR to test their newly developed lateral control system design change and verify that it corrected the hinge moment reversal/ uncommanded aileron deflection problem. These design changes, if effective, should be implemented on all new and existing ATR airplanes.

The board recommended that all ATR 42 and 72 aircraft be prohibited from flying in known icing conditions until the effects of upper wing ice buildup on the flying qualities and aileron hinge moment effects can be determined.

BEA Response

The BEA needed an entire second volume to contain their comments and disagreement with the NTSB's fundamental conclusion about the Roselawn accident [2]. Its response to the NTSB, the BEA expressed disappointment at their lack of participation in the investigation phase, findings, causes and safety recommendations. The BEA believed the NTSB had committed to allowing a strong contribution from the BEA in all phases of the investigation. The probable cause as originally released in draft form is shown below.

> *The National Transportation Safety Board determines that the probable causes of this accident were the loss of control, attributed to a sudden and unexpected aileron hinge moment reversal, that occurred after a ridge of ice accreted beyond the deice boots because: 1) ATR failed to completely disclose to operators, and incorporate in the ATR 72 airplane flight manual, flightcrew operating manual and flightcrew training programs, adequate information concerning previously known effects of freezing precipitation on the stability and control characteristics, autopilot and related operational procedures when the ATR 72 was operated in such conditions.*

The final probable cause statement, incorporating the strong sentiments provided by BEA, reads as follows:

> *The National Transportation Safety Board determines that the probable cause of this accident was the loss of control, attributed to a sudden and unexpected aileron hinge moment reversal, that occurred after a ridge of ice accreted beyond the deice boots while the airplane was in a holding pattern during which it intermittently encountered supercooled cloud and drizzle/rain drops, the size and water content of which exceeded those described in the icing certification envelope. The airplane was susceptible to this loss of control, and the crew was unable to recover.*

While the FAA pinned most of the problem on the airplane and its propensity to lose control when ice built up downstream of the deice boots, BEA sought to include the flight crew in their fault analysis. BEA claimed that the flight crew failed to comply with ATR procedures and that was a strong contributor to the accident. The fact that the FAA immediately banned both ATR models (42 and 72) from flying in icing conditions immediately after the accident, and restricted their flight envelope until ATR had redesigned, recertified, and installed new, larger deicing boots, indicates that the FAA saw the accident cause in a different light [3].

References

1. NTSB Aircraft Accident Report NTSB/AAR 96/01 7/9/96 Volume 1: Safety Board Report
2. NTSB Aircraft Accident Report NTSB/AAR-96/02 7/9/96 Volume II: Response of Bureau Enquetes-Accidents to Safety Board's Draft Report
3. Aviation Week & Space Technology, " ATR Boots pass Ice Test", May 1, 1995

Chapter 16
Conclusion

The commercial aviation industry has seen many improvements over the decades since its inception in the beginning of the twentieth century. The early propeller, internal combustion engine driven craft have now evolved into jet-powered aircraft that can hold nearly 600 passengers. The IC engine powered propeller planes broke ground on pressurized fuselages, which allowed planes to climb above the worst weather (except for takeoff and landing) and provide their passengers comfort and relief from the most severe turbulence.

The transports that now ply our skies have also evolved from the loud, fuel hogging turbojet planes that debuted in the 1950s to the new turbofan-powered planes that most airlines are clamoring for. However, the transition from propeller driven passenger craft like the DC-6 to the new A321 have required changes in how the airplanes are built, and what subsystems are employed to support the overall operation of the airplane.

While the first chapter provided an overall history of transport (i.e., passenger) aircraft, it only touched on the changes that came about in the transition from propeller to jet power. It did examine the changes that had been introduced during the first half-century of airplane travel that provided marked improvement in passenger safety and comfort.

This book details some of the systems that have been conceived and evolved over the decades to make air travel safe while traveling across the globe at over 500 mph. It also details some of the tragedy that has occurred when designers, maintenance personnel, and sometimes even the flight crew failed.

When examining this recent aviation history, it is important to remember the roles of the various federal agencies that interact with air travelers. In the USA, the FAA sets safety standards for the design, fabrication, and operation of these airplanes. However, when they fail, it is the National Transportation Safety Board which is charged with investigating every civil aviation accident in the USA. The NTSB also has responsibility for advocating and promoting new safety recommendations, frequently at the conclusion of their accident investigation. The NTSB traces its roots to the Air Commerce Act of 1926, when a congressional act mandated

© Springer Nature Switzerland AG 2020
T. Filburn, *Commercial Aviation in the Jet Era and the Systems that Make it Possible*, https://doi.org/10.1007/978-3-030-20111-1_16

that the Commerce Dept investigate all aircraft accidents. However, it is the FAA's role whether to accept these recommendations and mandate them for the existing fleet and/or introduce these changes into new aircraft. The international equivalent of the FAA is the European Aviation Safety Agency, while the Bureau of Enquiry and Analysis for Civil Aviation Safety (BEA) is an agency of the French Government that investigates aviation accidents in France, or through agreements with foreign governments accidents involving aircraft designed and/or built in France (e.g., Airbus, ATR).

Chapter 2 described the present generation of systems that have evolved to support the higher speeds, higher wing loading, and larger aircraft now operating around the world with the major airlines. These changes included improvements necessitated by the higher speeds desired and offered by jet power. With airplane drag being a function of velocity squared, the power required to operate planes at M = 0.82 (530 mph) vs. the top end propeller transport (M = 0.5, 330 mph) is a roughly 2.7 × increase. While the turbojet and now turbofan engines were able to increase the engine power available (in a reasonable volume) to achieve these new velocities, they required changes in the aircraft systems.

Wing design evolved between the propeller and turbine era. A quick look at the DC-7 wing shows a lifting surface perpendicular to the fuselage out to the tip of the wing. The higher speeds, near sonic, now being flown by the jet-powered planes meant a new wing configuration. The Boeing 707 and all the following jet-powered craft used a swept wing design to reduce drag and increase stability at the high speeds now being flown. These wings were optimized for the high-speed cruise phase of their missions that comprised >90% of their flight time. It made takeoff and landing more difficult. These wings were designed to provide significant lift when operating at M = 0.82, they did not provide nearly enough lift when the airplane was operating at reasonable takeoff and landing speeds. A reasonable takeoff speed can be defined by the runway length required to accelerate the airplane to takeoff speed. Similarly a reasonable landing speed can be defined as the speed that can be stopped in the distance of a normal runway, without excessive G forces imposed on the passengers (e.g., the landing gear arresting wire of a US Navy carrier would not work). Figure 16.1 shows the change in wing drag coefficient between a straight wing (e.g., DC-7) and a swept wing (e.g., A320). While the drag coefficients are both low below M-0.7, there is a significantly higher drag coefficient for the straight wing at the new higher speeds (M = 0.82) of the jet aircraft.

The military with even higher top speeds attempted to overcome this lift vs. drag conundrum by developing moving wing geometry. The F-111 and the F-14 both took off and landed with wings in a conventional straight configuration, however they could be adjusted in flight to a swept arrangement to allow reasonable drag at higher (>M = 1) speeds. Commercial airliners could not afford the cost nor the complexity of moveable wings. The compromise reached for commercial jet airliners relied on moveable flaps (wing trailing edge surfaces) and slats (wing leading edge surfaces) to change the lift characteristics of the wing. These moveable devices get employed during takeoff and landing to greatly increase the lift (but also drag) of the wing. These enhancements allow commercial airliners to takeoff with reasonable

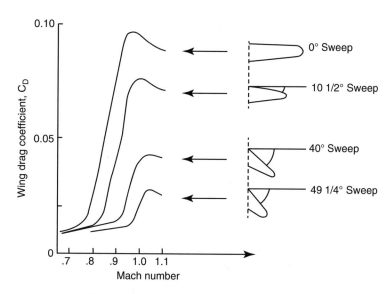

Fig. 16.1 Swept vs. straight wing drag coefficient [1]

takeoff runs and land on runways of less than 8000 ft. These same flight control surfaces get stowed during cruise, allowing a low-drag aerodynamic shape to be employed.

A second critical improvement that has accompanied these flight control surfaces has been the use of high-pressure hydraulic systems to move them during flight. The high speeds now seen by these large commercial aircraft will generate enormous aerodynamic loads opposing the movement of any flight control surface into the airstream. High-pressure hydraulic systems boost the power available to move and control these surfaces well beyond the strength that could be generated by a pilot's muscle.

Chapter 9 details what happens when the cockpit no longer can make use of its flight control surfaces. This chapter details the extraordinary efforts that the flight crew operating United 232 made to land their aircraft after an engine failure produced a total loss of all the hydraulic systems on-board. These pilots brought their DC-10 into a controlled landing at Sioux City, IA, simply through changes in engine power of the two remaining wing-mounted turbofan engines. This controlled landing occurred despite the total loss of hydraulic pressure and thereby complete inability to move any control surfaces (including high lift, rudder, elevator, and aileron). The aircraft was a complete loss after breaking up upon landing but 185 of the 296 persons on-board survived the accident.

Chapter 3 discusses the dramatic changes that have occurred in the area of airplane propulsion. Propellers were the devices that provided thrust from the first days of the Wright brothers through the 1940s. However, the development of the gas turbine engine during WWII ushered in a new propulsion technology. These

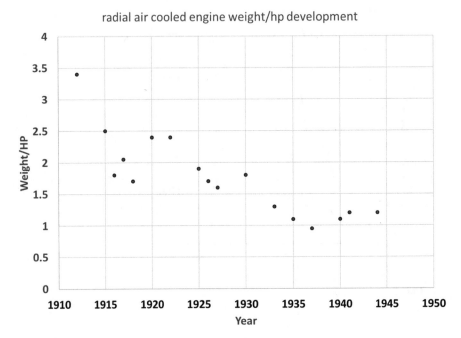

Fig. 16.2 Radial air cooled engine weight/hp development

powerplants provided a much higher thrust capability in a lighter weight vs. the internal combustion engine/ propeller combination they replaced. Figure 16.2 shows the trend in weight per HP for air cooled piston engines and that it had leveled at about 1 lb/HP near the end of WWII [2]. As previously shown in Figure 3.1 turbofan engines have continued to improve on that figure.

Chapter 4 documents the history and development of both the cabin pressurization and air management system (AMS). These coupled systems regularly (but not always) keep a comfortable temperature within the cabin, while also providing a pressurized (above outside ambient) environment. These AMS routinely uses an air cycle system to provide heat and cooling for the passenger space conditioned air. The introduction and development of pressurization and AMS were closely coupled. Both these systems provided for passenger comfort against both the vagaries of weather, turbulence and temperature extremes found at altitude. The metabolic requirement for oxygen limits the altitude at which humans can perform or even stay conscious. However, weather systems and higher levels of turbulence can frequently be found at elevations up to 30,000 ft. (higher than Mt. Everest), thereby insuring that any aircraft that went above the turbulence would exceed human tolerance. So, to maintain pilot, crew, and passenger cognition, airplanes rely on pressurization systems to raise the internal pressure to an equivalent lower altitude. The pressure that airplanes reach during their flight becomes a compromise between passenger comfort (lower altitude, higher pressure) and fuselage thickness to retain that pressure (thicker walls equals greater weight, fewer passengers). Airbus and

Boeing (plus the commuter airline manufacturers) have generally settled on cabin equivalent altitudes of 6000–8000 ft. altitude, lower than sea level pressure (hence your ears may pop going up or down), but well within the tolerance of the human respiratory system.

Chapter 10 details an accident that occurred on a United flight (811) leaving Honolulu bound for Sydney. The same cabin pressurization system that keeps passengers and crew alert during the flight creates a significant pressure difference between the inside cabin environment and the outside air. At 40,000 ft. altitude, with an equivalent 6000 ft. altitude inside the cabin, this pressure difference can reach over 9 lbf over every square inch of exposed surface.

Chapter 4 detailed how this pressure difference must be resisted over numerous cycles (take off to cruise and landing) and demonstrated how one of the earliest jet transports (De Haviland Comte) did not design well for the metal fatigue that occurs over these multiple pressurize/depress cycles. United airlines flight 811, a Boeing 747, also had a design flaw from its initial entry into service. The new 747 provided a large increase in passenger capacity, but beneath the cabin floor, it also created a much larger cargo volume. The aircraft designers wanted to take advantage of that cargo capacity, without impacting the passenger experience. Therefore, they designed a cargo door that could open outward to allow full access to the interior cargo compartment.

The Boeing 747 cargo door was a key component in keeping the fuselage and all occupants at the higher pressure desired when at altitude. Most passenger doors open inwardly initially, which means they are kept sealed and in place by the cabin pressure differential (higher inside vs. outside), the cargo door would have to resist the pressure difference directly through its attaching and locking features. This door had been designed with multiple latches and cams to link with the fuselage at eight points along the bottom. The hinge attached to the fuselage at the top provided the upper restraint mechanism. The door also had a mid-span latch on either side, about ½ way down the length of the door.

Unfortunately for the passengers on United flight 811, the balky cargo door had shown problems in the past and in fact had forced a 747 bound for New York to return to London when it experienced pressurization problems while climbing through 20,000 ft. An examination of the door upon landing showed a 1 ½ inch gap along the bottom of the door, but the cargo warning light in the cockpit did not annunciate. When the door explosively let go on Flight 811, it buckled the main cabin floor and sent 10 first class seats out of the plane, along with 9 passengers. The pilot was able to safely return the aircraft and remaining passengers to Honolulu. This accident underscored the large pressure difference found in all modern airliners traveling at altitude. In addition, it demonstrates the energy that can be released when that high-pressure air (relative to ambient) explosively leaves the aircraft.

Chapter 5 examined those systems that get used for less than 1% of the flight, the landing gear and their individual wheels and brakes. The proper operation of the landing gear plus the wheels and brakes are imperative for the safety of the passengers. The massive weights that new aircraft are reaching, coupled with the inherent sink rate required to descend to the airport means that the landing gear absorb an enormous load. That continuous load plus the instantaneous energy absorbing shock of touching down must both be supported by the landing gear.

The wheels and brakes have similar brief but large load requirements. The wheels go from stationary to landing speed in a fraction of a second (hence the large skid pad and smoke when first touching down). The brakes must absorb MJ of energy in less than 10 s and turn the plane's kinetic energy into thermal energy (heat up). Finally, all of these systems must work on snowy, wet runways, when anti-lock braking systems may be limiting the stopping power of the brakes.

Chapter 12 details the Concorde accident that occurred upon takeoff from Paris' Charles De Gaulle airport. This supersonic transport had a delta wing shape, required for low drag necessary at its supersonic cruise speed. This same delta wing shape could not benefit from the same high-lift devices of the more conventional swept wing. Instead, it required a very high takeoff velocity than nose rotation to get lift. The high takeoff velocity meant higher wheel loads, higher wheel speeds, and unfortunately higher susceptibility to runway foreign object damage (FOD, anything that didn't belong on the runway). A French Airways Concorde tire hit a piece of titanium that had just fallen off a Continental Airlines DC-10.

The Concorde ran over the DC-10 titanium strip at the worst point during its takeoff acceleration run. By the time the impact of the strip was noticed, the Concorde was traveling too fast to abort the takeoff. However, the damage initiated by the DC-10 part breached a fuel tank, produced a loss of thrust on 2 engines (from the fuel leaking and burning near the engines and spreading to the wing fuel tank), and prevented the gear from retracting. The combination of the low speed inherent in takeoff, the loss of thrust from 2 engines, and the added drag of the landing gear remaining extended made the aircraft extremely unstable. It crashed into a hotel near the airport with a loss of everyone on-board.

Chapter 6 discusses the design and operation of the fuel storage and transfer systems required for large transport aircraft. The long-range design of today's intercontinental airplanes coupled with high passenger loads (the A380 can hold over 600 passengers) mean that large quantities of fuel must be stored and then transferred to the engines during flight. In addition, the fuel must also be frequently transferred during flight to keep the Cg in an appropriate point on the aircraft (both fore and aft, plus port and starboard).

Chapter 13 examines what happened with TWA 800, a Boeing 747 that exploded shortly after takeoff from New York's JFK airport. The investigation into the explosion ultimately placed the blame on an unintended and unexpected electrical anomaly inside the center wing fuel tank (CWT). This CWT had both an extremely low level of liquid fuel inside it, meaning the vast majority of the tank was filled with fuel vapor. In addition, this same CWT had received significant heat input due to its location (directly above the air management system, AMS) and the operation of the aircraft's AMS for several hours before takeoff. While TWA 800 was a complete loss including all crew and passengers, it did introduce changes into the fuel systems. By 2005, the FAA mandated that all CWTs be provided with a nonexplosive atmosphere. In reducing this mandate to practice, the airplane manufacturers have elected to rely on nitrogen enriched air generation systems. These systems take

pressurized and based on varying gas permeabilities through a membrane wall produce a gas stream that has less than 12% oxygen, sufficient to make the ullage volume a nonexplosive environment.

Chapter 7 details the seemingly enormous change in instruments, sensors, and the human-machine interface (HMI). While it is true that the cockpit environment has changed dramatically from the open-air days of the Wright brothers, the fundamental instrument for measuring airspeed (pitot probe) has not really changed since the 1940s. How this data gets presented in the cockpit has improved, with new digital cockpit displays placing a wealth of data in the hands of the flight crew.

Chapter 14 explains what can happen to a modern jet airliner when one of those pitot probes fails to provide accurate speed data. This Airbus A330 suffered an icing incident on one of pitot probes that was feeding airspeed information into the autopilot system. When the probe iced over due to heavy icing conditions, the autopilot automatically disconnected, and the flight crew responded poorly. They were not well trained in how to diagnose nor recover from an aircraft stall incident at high altitude (when the plane had little margin for climbing). The resultant accident caused the death of all 228 souls on-board.

Chapter 8 describes the long history and premature announcement of success in defeating ice as a hazard to flight. This chapter identifies one of the primary initiatives of the National Advisory Committee for Aeronautics (NACA, the precursor to NASA) which started in the 1920s. One of its major efforts was aimed at reducing the hazard posed by airplane icing. This group helped to develop the early pneumatic deicing systems that helped transport aircraft in the 1930s. NACA performed testing to validate the use of engine exhaust heat as a viable wing, leading edge deicing system. This thermally focused system has now evolved into using high-temperature compressed air from the engine, before it has entered the combustor or turbine sections, as the deicing system on large-scale transport aircraft.

Chapter 15 discusses the Roselawn accident, where an ATR 72, turboprop, commuter aircraft was brought down by ice buildup on Halloween 1994. This accident highlighted the lack of insight into the fact that even modern aircraft (this ATR 72 was built that same year 1994) could still be susceptible to icing conditions. This accident also displayed a rift between the US NTSB and the French BEA. The NTSB highlighted previous role problems with this aircraft when operating in heavy icing conditions, with the flaps set at 15°. In addition, the NTSB claimed that inadequate information had been offered regarding the performance of the ATR aircraft during this type of icing incident. The BEA tried to deflect this aircraft design criticism and offered that the flight crew failed to follow standard procedure. This accident did highlight the lack of understanding to the magnitude of icing that could be produced by various meteorological conditions including supercooled large liquid droplets.

All of these chapters have painted a picture of the complex nature of jet travel in the twenty-first century. While these systems have evolved to make airplane travel the safest form of travel today (as measured by passenger-mile), it also demonstrates the frailty of the overall system if one of these subsystems becomes compromised.

References

1. https://history.nasa.gov/SP-367/contents.htm   https://history.nasa.gov/SP-367/contents.htm, retrieved 11/25/18
2. The Engines of Pratt & Whitney: A Technical History, Connors, J., AIAA, 2010

Index

© Springer Nature Switzerland AG 2020
T. Filburn, *Commercial Aviation in the Jet Era and the Systems that Make it
Possible*, https://doi.org/10.1007/978-3-030-20111-1

Printed in the United States
By Bookmasters